Conference Proceedings of the Society for Experimental Mechanics Series

Series Editor
Tom Proulx
Society for Experimental Mechanics, Inc.,
Bethel, CT, USA

For further volumes:
http://www.springer.com/series/8922

Alvaro Cunha
Editor

Topics in Dynamics of Bridges, Volume 3

Proceedings of the 31st IMAC, A Conference on Structural Dynamics, 2013

Springer

Editor
Alvaro Cunha
Department of Civil Engineering
Universidade do Porto
Praça Gomes Teixeira
Porto, Portugal

ISSN 2191-5644 ISSN 2191-5652 (electronic)
ISBN 978-1-4614-6518-8 ISBN 978-1-4614-6519-5 (eBook)
DOI 10.1007/978-1-4614-6519-5
Springer New York Heidelberg Dordrecht London

Library of Congress Control Number: 2013932014

© The Society for Experimental Mechanics, Inc. 2013
This work is subject to copyright. All rights are reserved by the Publisher, whether the whole or part of the material is concerned, specifically the rights of translation, reprinting, reuse of illustrations, recitation, broadcasting, reproduction on microfilms or in any other physical way, and transmission or information storage and retrieval, electronic adaptation, computer software, or by similar or dissimilar methodology now known or hereafter developed. Exempted from this legal reservation are brief excerpts in connection with reviews or scholarly analysis or material supplied specifically for the purpose of being entered and executed on a computer system, for exclusive use by the purchaser of the work. Duplication of this publication or parts thereof is permitted only under the provisions of the Copyright Law of the Publisher's location, in its current version, and permission for use must always be obtained from Springer. Permissions for use may be obtained through RightsLink at the Copyright Clearance Center. Violations are liable to prosecution under the respective Copyright Law.
The use of general descriptive names, registered names, trademarks, service marks, etc. in this publication does not imply, even in the absence of a specific statement, that such names are exempt from the relevant protective laws and regulations and therefore free for general use.
While the advice and information in this book are believed to be true and accurate at the date of publication, neither the authors nor the editors nor the publisher can accept any legal responsibility for any errors or omissions that may be made. The publisher makes no warranty, express or implied, with respect to the material contained herein.

Printed on acid-free paper

Springer is part of Springer Science+Business Media (www.springer.com)

Preface

Topics in Dynamics of Bridges represents one of seven volumes of technical papers presented at the 31st IMAC, A Conference and Exposition on Structural Dynamics, 2013, organized by the Society for Experimental Mechanics, and held in Garden Grove, California, February 11–14, 2013. The full proceedings also include volumes on Nonlinear Dynamics; Experimental Dynamics Substructuring; Dynamics of Civil Structures; Model Validation and Uncertainty Quantification; Special Topics in Structural Dynamics; and Modal Analysis.

Each collection presents early findings from experimental and computational investigations on an important area within Structural Dynamics. The Dynamics of Bridges is one of these areas.

This focus area seeks to provide a dedicated forum for researchers, designers, and bridge owners to exchange ideas and concepts relating to the development, implementation, and application of efficient experimental and numerical tools for the assessment, monitoring, and control of the dynamic behavior of roadway, railway, or pedestrian bridges, taking in mind problems of human comfort, structural performance and durability, damage detection and location, or structural rehabilitation. Particular attention is devoted to innovative applications in full-scale structures.

The organizers would like to thank the authors, presenters, session organizers, and session chairs for their participation in this track.

Porto, Portugal

Alvaro Cunha

Contents

Chapter 1
Vibration Analysis and Structural Identification of a Seismically Isolated Bridge

C. Bedon and A. Morassi

Abstract This work is part of a current project developed in collaboration with the Local Highway Authority of the Friuli Venezia Giulia, a region located in the North East of Italy. Aim of the project is to define a periodic structural-condition monitoring of bridge structures based on dynamic measurements to ensure that they provide a continued and safe service. The main objective of this paper is to present the results of structural identification based on dynamic data on a two-span post-tensioned reinforced concrete bridge recently built in an area of high seismicity. The span length is 37.5 m and the deck is simply supported by two multi-directional elastomeric seismic bearings on each abutment and on the central pier. Harmonic forced vibration tests with vertical, transversal and longitudinal direction of the excitation were carried out to determine the dynamic parameters of the lower vibration modes of the bridge. An identification procedure based on modal analysis and finite element model updating is presented for the characterization of the bridge and, particularly, for the identification of the stiffness of the bearing devices supporting the deck superstructure.

Keywords Base-isolated bridges • Dynamic tests • Finite-element modeling • Modal analysis • Structural identification

1.1 Introduction

Seismic isolation has become a fairly common technique to protect bridges from severe earthquakes. The use of this technique was initially confined to large bridges. More recently, base isolation solution has been applied even to small and medium-side bridges. This is the result of the imposition of more severe national seismic codes (such as in Italy, for example) and of refined seismic zoning.

In this paper the results of a campaign of harmonically forced vibration tests conducted on a two-span post-tensioned reinforced concrete bridge with deck supported on six elastomeric isolators are presented and discussed. The bridge is built in an area of the Friuli Venezia Giulia Region (Northern Italy) with a high level of seismic activity.

One of the purposes of the investigation was to verify the reliability of numerical models to describe the linear dynamic behavior of the bridge. It is well-known, in fact, that dynamic data (namely, natural frequencies, mode shapes, and damping factors) can provide meaningful results if they are used to improve a finite element model of the bridge, enabling, for example, to estimate mechanical properties of the materials and description of boundary conditions [1–7]. Incidentally, a review of the literature in the field reveals that experimental research works on the dynamic behavior of isolated bridges of this typology are few, see, for example, [8, 9].

Another important purpose of the present research was the definition of a baseline model of the bridge for future investigation of diagnostic character. This issue is of great importance for the company who manages the local highway network. In fact, a long-term plan of maintenance of the bridge cannot neglect possible changes of performances experienced by the bearing devices. One possible way to check the functionality of these devices is the periodic removal of isolators from bridge for a laboratory test campaign. This appears a not feasible approach, due to high economic impact, particularly since

C. Bedon
DICAR-University of Trieste, Trieste, Italy

A. Morassi (✉)
DICA-University of Udine, Udine, Italy
e-mail: antonino.morassi@uniud.it

A. Cunha (ed.), *Topics in Dynamics of Bridges, Volume 3: Proceedings of the 31st IMAC, A Conference on Structural Dynamics, 2013*, Conference Proceedings of the Society for Experimental Mechanics Series 38,
DOI 10.1007/978-1-4614-6519-5_1, © The Society for Experimental Mechanics, Inc. 2013

traffic interruption is required. A more reliable approach is to test periodically the response of the bridge with devices in service, to localize and quantify global response changes as generated by local degradation of conventional structural elements as well as by isolation devices. Most of the diagnostic methods for damage detection in bridges, in fact, are based on a comparison between the structural response of the actual – possibly damaged – state and that associate to a reference (say, undamaged) configuration of the bridge [10–12]. It follows that, in order to be able to give a proper interpretation of the damage-induced changes on the dynamic characteristics of the construction, it is crucial to have at disposal an accurate knowledge of the undamaged configuration of the bridge, and this was ultimately the main goal of this study.

Harmonic forced vibration tests with vertical, transversal and longitudinal direction of the excitation were carried out to determine the dynamic parameters of the lower vibration modes of the bridge. The main results of an identification procedure based on modal analysis and finite element model updating are presented for the characterization of the bridge and, particularly, for the identification of the mechanical properties of the bearing devices supporting the deck superstructure.

1.2 Bridge Description

The main geometrical features and dimensions of Dogna Bridge are shown in Fig. 1.1. The structure is a two-span, two-lane post-tensioned reinforced concrete (RC) bridge. The span length is 37.5 m and a single pier is present at the middle cross-section. The abutments consist of vertical RC walls. The deck is continuous at the inner support on the pier and it is simply supported by two multi-directional elastomeric seismic bearings on each abutment and on the pier, see Fig. 1.2. Pier and abutments were built on cast-in-place concrete piles. The typical cross-section of the bridge is formed by a central rectangle,

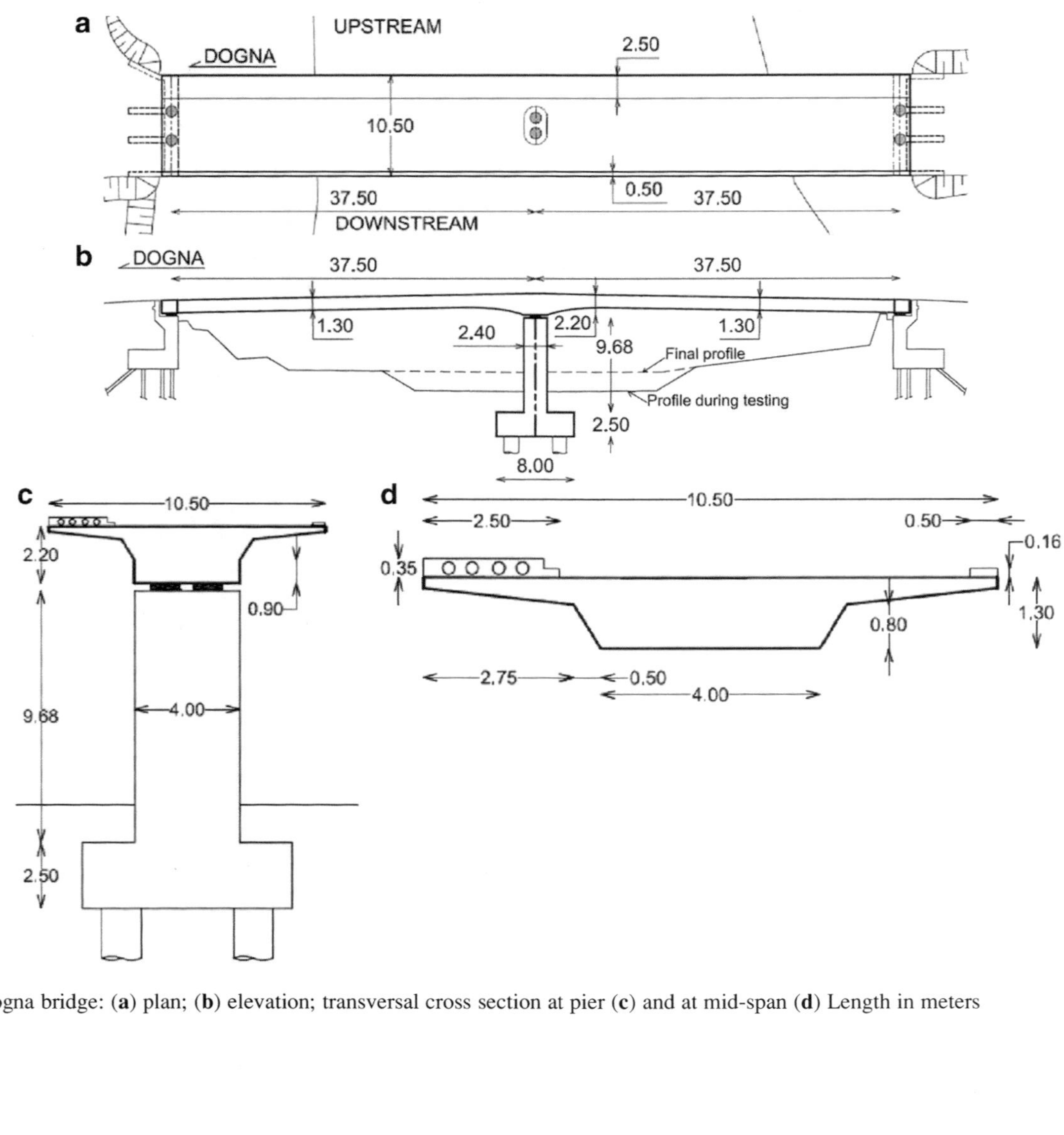

Fig. 1.1 Dogna bridge: (**a**) plan; (**b**) elevation; transversal cross section at pier (**c**) and at mid-span (**d**) Length in meters

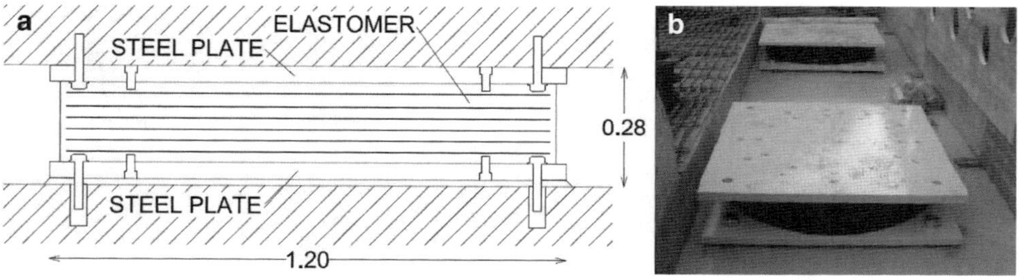

Fig. 1.2 Elastomeric bearings of Dogna bridge: (**a**) vertical cross-section; (**b**) detail. Length in meters

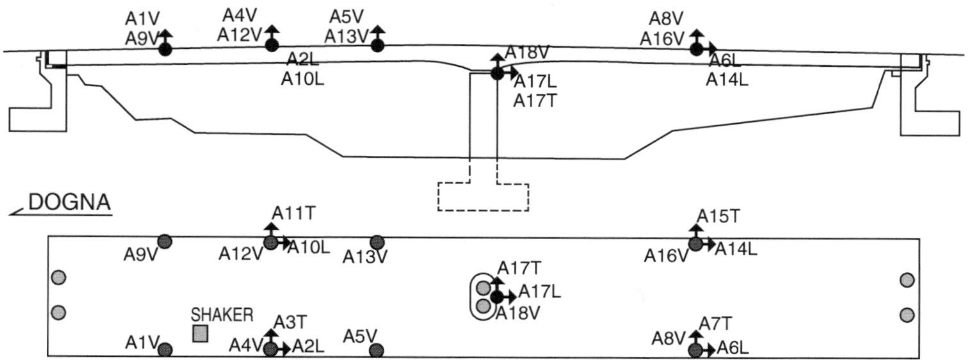

Fig. 1.3 Instrumental layout

approximately 5.0 m in width and 1.3 m in thickness, which contains the pre-stressing tendons, and two lateral portions with thickness varying from 0.5 to 0.2 m. The thickness of the central part of the deck cross-section gradually increases from mid-span toward the pier support. The construction of the bridge was completed in the Spring of 2007. It was designed under the Italian Standard Specifications for Highway Bridges in Seismic Areas. Dynamic tests were carried out in April and May 2007, respectively.

1.3 Dynamic Testing: Results and Comments

In May 2007 the bridge underwent an extensive series of harmonic forced-vibration tests with low levels of excitation. Steady-state harmonic vibrations were induced on the bridge by means of the stepped-sine technique in order to measure the frequency response function (FRF) in the range from 1 to 15 Hz. The maximum amplitude of the force generated by the mechanical exciter during the tests was equal to 20 kN. The exciter was mounted on the bridge deck approximately at one third of the span of the Dogna side, near the downstream sidewalk, see Fig. 1.3.

Structural response was simultaneously measured by 18 seismic accelerometers ICP (Series 393B12), having sensitivity equal to 10 V/g. Based on the experimental layout shown in Fig. 1.3, the response of the bridge was separately measured under excitation parallel to the vertical (V), longitudinal (L) and transverse (T) direction. The tests were carried out in absence of the asphalt layer on the deck. FRF measurements were carried out with a frequency step increment of 0.05 Hz. A smaller step, equal to 0.02 Hz, was adopted in the two bands (1.8, 3.4) Hz, (6.5, 8.5) Hz in order to improve the identification of vibration modes under vertical excitation. A data acquisition unit was used for data acquisition, control of the force actuator, and automated signal processing.

Experimental Modal Analysis (EMA) techniques based on multiple curve-fitting have been applied to estimate the dynamic parameters of the bridge, namely natural frequencies, damping ratios and related mode shapes. A careful analysis of experimental FRF curves suggested the use of responses to excitation parallel to the transversal (T), longitudinal (L) or vertical (V) direction for the reconstruction of the dynamic properties of the vibrating modes with prevailing amplitudes along the T, L or V-direction, respectively. The experimental FRF curves are rather regular in the low frequency range, say from 1 to 6 Hz. Some difficulties have been encountered in the analysis of the FRFs at high frequencies, especially from 12 to 15 Hz, for the presence of a more pronounced irregularity and low amplitude levels. In these cases, it was appropriate to

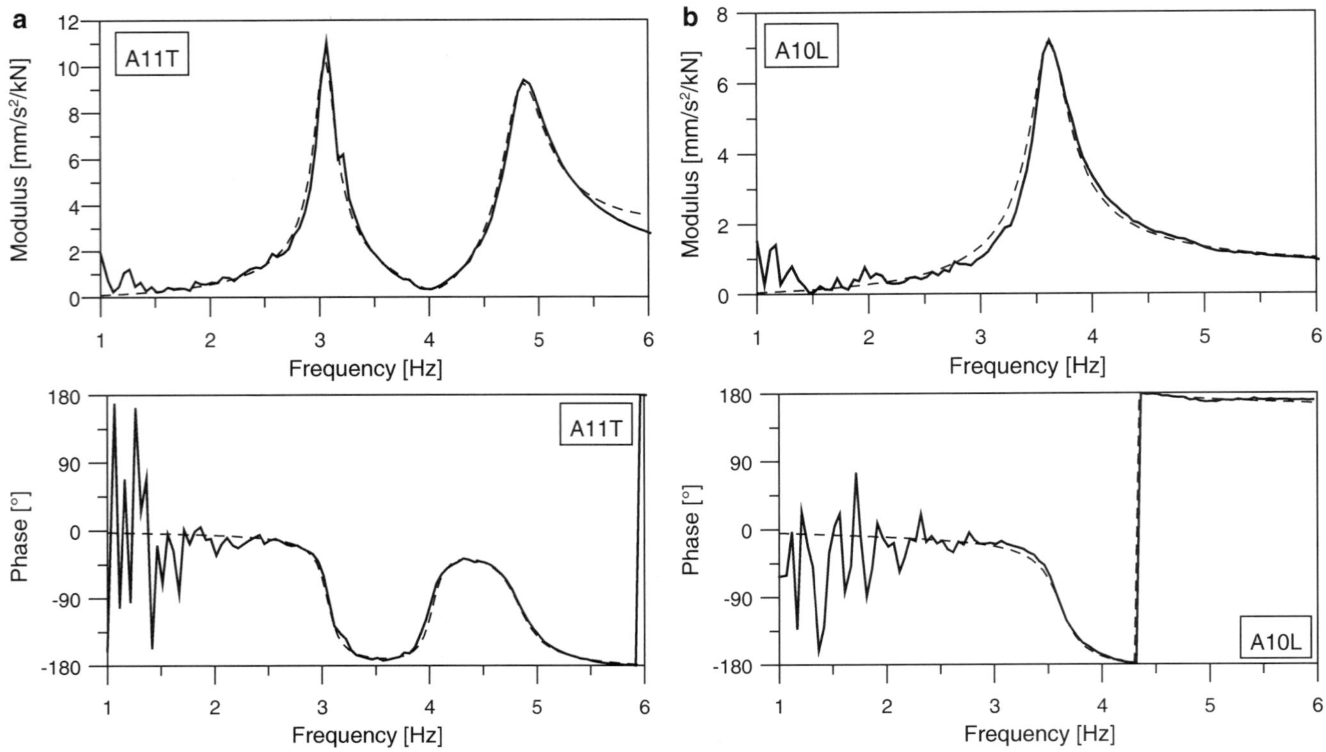

Fig. 1.4 Example comparisons between measured (*continuous line*) and synthesized (*dashed line*) point inertance for Dogna bridge: (**a**) accelerometer A11T; (**b**) accelerometer A10L

Table 1.1 Experimental Modal Analysis results: mean value of natural frequencies p_r and damping ratios ξ_r, with their maximum deviations. T = Torsional; B = Bending; RB = (almost) rigid-body motion

Mode order r –	Description –	Natural frequency p_r [Hz]	Damping ratio ξ_r [%]
1	1st B	2.022 ± 0.001	0.88 ± 0.03
2	RB Transversal	3.053 ± 0.003	2.88 ± 0.05
3	2nd B	3.180 ± 0.002	0.89 ± 0.05
4	RB Longitudinal	3.605 ± 0.002	4.33 ± 0.07
5	RB Torsional	4.831 ± 0.011	3.93 ± 0.13
6	3rd B	6.887 ± 0.046	Not available
7	1st T	6.934 ± 0.015	1.15 ± 0.20
8	2nd T	7.995 ± 0.005	0.88 ± 0.10
9	4th B	9.107 ± 0.020	1.78 ± 0.44
10	Coupled B-T	12.910 ± 0.025	1.66 ± 0.20
11	Coupled B-T	14.228 ± 0.081	0.66 ± 0.12
12	Coupled B-T	14.433 ± 0.100	0.77 ± 0.27

repeat several curve-fitting runs by varying both the modal order and the frequency interval. Figure 1.4 shows a comparison between some measured and synthesized point inertance function of the bridge.

Vibration modes have resonance frequencies well separated, with the exception of the pairs of modes (2,3) and (6,7). The identification of Modes 2 and 3 was made easy by the fact that their modal shape corresponds to vibrations with prevailing amplitudes in the horizontal plane and along the vertical direction, respectively. Modes 6 and 7 are both primarily vertical modes and, in order to identify them, it is found convenient to analyze the FRFs obtained as the half-difference and the half-sum of the FRFs measured in points belonging to the same cross-section of the deck in order to separate the flexural and torsional contribution on these modes. Table 1.1 summarizes the results of the EMA.

Deviations of the natural frequency values from their average value can be considered negligible for modes at low frequency and, generally, differences increase with the mode order. The estimation of the damping ratios is equally good, even if, as expected, differences are important for some modes, particularly for higher order modes. Moreover, mode shapes

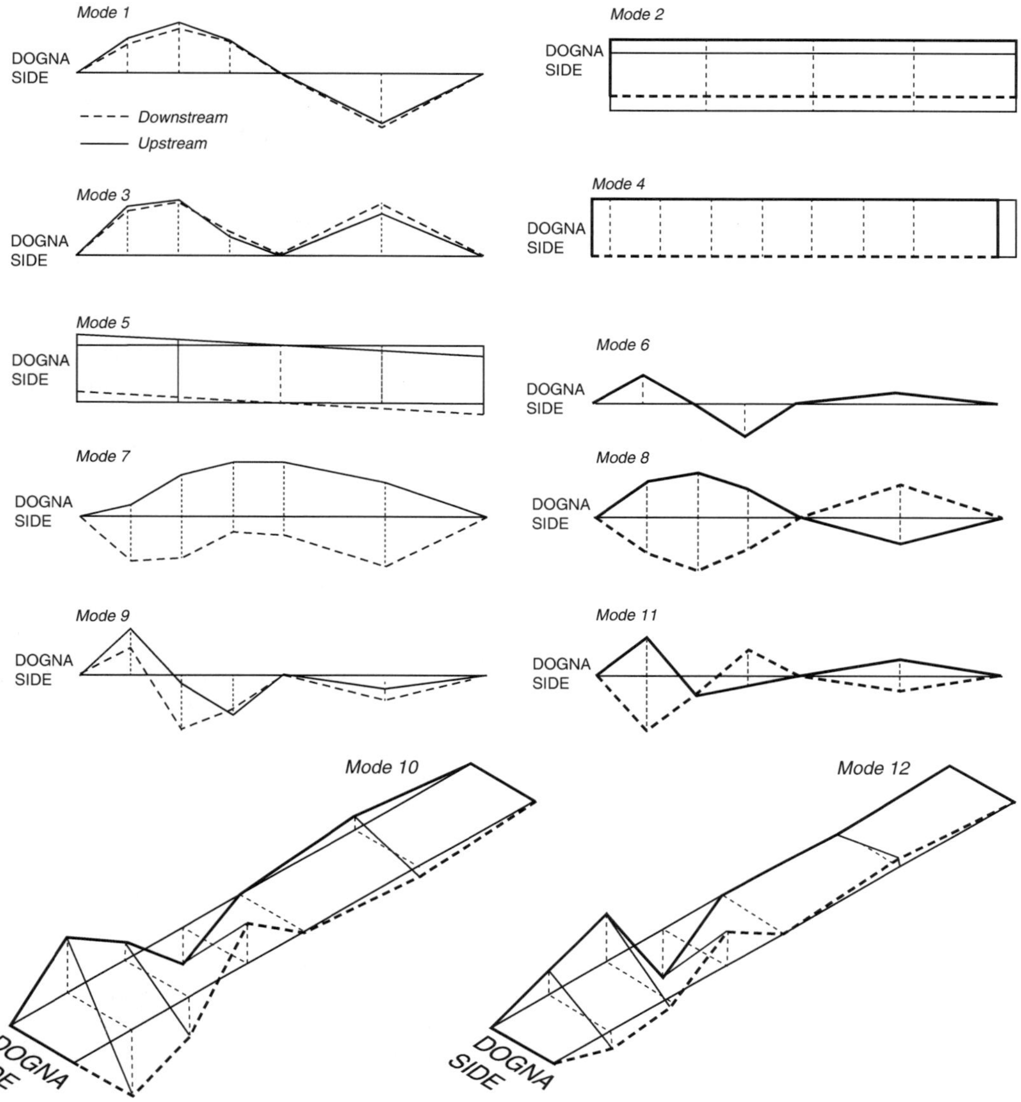

Fig. 1.5 3D view of first 12 experimental mode shapes

components can be considered real-valued with good approximation. Mode shapes in Fig. 1.5 correspond to linear interpolation of measured modal components and are normalized to the maximum amplitude component.

The presence of the elastomeric supports on abutments and pier produces a decoupling of the dynamic response of the bridge. In fact, the experimental vibration modes (EMA modes) are grouped into two classes: the vibration modes with modal shape corresponding to a (quasi) rigid-body deformation of the whole deck, and the vibration modes associated with flexural/torsional deformation of the deck in the vertical plane. The first class includes EMA Modes 2, 4 and 5, see Table 1.1. Mode 1 (2.02 Hz) corresponds to the fundamental bending mode of a two-span beam. The torsional effects of the deck are almost absent and the vertical modal components evaluated on a pair of transversally aligned points differ from each other by about 10 %, with amplitudes of oscillation that are bigger on the upstream side. The two spans show in-phase vibrations in Mode 3 (3.18 Hz), with a spatial shape which is similar to the second bending mode of a simply supported continuous two-span beam. A loss of symmetry in the longitudinal direction emerged in this mode, with amplitude of vibration at mid-span of Dogna side which is about 20 % bigger than the corresponding amplitude measured on the other span. Mode 6 (6.89 Hz) is reminiscent of the third bending mode of a two-span continuous beam, even if the sparsity of the measurement points on the left span makes difficult the reconstruction of the complete mode shape. Next two modes, Mode 7 (6.93 Hz) and Mode 8 (8.00 Hz), are the first (fundamental) and second torsional modes of the bridge deck. A significant coupling between bending and torsional deformation of the deck is present in higher modes.

1.4 Interpretation of the Experiments and a Preliminary FE Model

The objective of the interpretation of dynamic tests was to determine a FE model of the bridge able to accurately predict its dynamic behavior within the frequency range of interest for seismic analysis. A 2D-version of this FE model is used in [13] to carry out a seismic analysis of the bridge.

A 3D Preliminary FE model was constructed based on the geometric details of the structure, see Fig. 1.6. The simulations are performed within the SAP2000 package. The FE model is created using solid FEs to model the deck and the pier. Typical FE dimension for deck structure is of order 0.25 and 0.50 m in the transverse and longitudinal direction of the bridge, respectively. The FE mesh has been refined to model the change in geometry of the deck near the central pier and near the lateral footway.

The inertial/stiffening effects of the footways were neglected in this model. The abutments are considered as non-deformable (rigid foundations), whereas the pier is assumed to be rigidly connected to the foundation, ignoring soil-structure interaction effects to the dynamics of the bridge system.

Concerning the effect of prestressing, many studies agree that it tends to decrease the natural frequencies. Other researches report that prestressing has no or negligible effect on the natural frequencies, see [14] for a recent account on this issue. Here, the prestress force effect has been neglected in the eigenvalue analysis.

Base on material tests carried out during construction, the Young modulus of the concrete Ec was assumed equal to 43.20 GPa, both for deck and pier. The mass density of concrete was 2,500 kg/m^3.

Bearing devices were modeled by 3D solid FEs, see Fig. 1.6. In order to ensure a perfect correspondence between nodes of the bearing devices and deck FE mesh, each isolator has been described by a parallelepiped with base dimension 1.20×1.00 m, and a mesh of 6×4 FEs in plan and 6 FEs in thickness has been introduced. Bearing devices are assumed to be made by an "equivalent" linearly elastic orthotropic material, with Young moduli E_x, E_y, shear moduli G_x, G_y, and Poisson coefficient $\nu = \nu_x = \nu_y$, where X and Y are the transverse and longitudinal directions of the bridge deck. The identification of effective elastic characteristics of the isolators represents a crucial point of identification. It is well known, in fact, that nominal value of the (global) stiffness supplied by the manufacturer generally is significantly less than the effective shearing stiffness under testing or ordinary traffic conditions. The reason for this large difference lies in the fact that the strain experienced by the bearing during testing was much lower than the one associated with the mechanical characterization under static or quasi-static load conditions in the laboratory, see, for example, [9].

A first mechanical characterization of the bearing devices, leading to the Preliminary FE Model of the bridge, has been carried out assuming rigid-body motion of the deck. Therefore, isolator stiffness K_x, K_y associated to transverse (X-direction) and longitudinal (Y-direction) shear displacements, have been determined by imposing a perfect matching between analytical and experimental natural frequency values associated to RB Transverse and RB Longitudinal modes. Considering a total mass of the deck equal to $M = 1{,}612$ t, the stiffness turn out to be equal to $K_x = 98.8$ MN/m, $K_y = 137.8$ MN/m. Based on these values, the Young and shear moduli of the material forming each isolator have been determined and insert into the FE model. Note that the nominal value of the shearing stiffness of the isolator is $K^{nom} = 71.25$ MN/m. The axial stiffness of the bearing was taken coincident with the nominal value $K_z = 7631$ MN/m.

Table 1.2 lists the values of the natural frequencies predicted by the Preliminary FE model of the bridge. A direct comparison between vibration modes obtained experimentally and analytically revealed that all the measured mode shapes have their counterparts in the Preliminary FE model, with the exception of EMA Modes 10 and 12. As it can be seen in Table 1.2, the order of the sequence is partially changed since EMA Modes 6,7,8,11 have correspondence to the set of analytical Modes 7,8,10,16, respectively. The lack of correspondence to FEA Mode 6, which is the first horizontal "elastic"

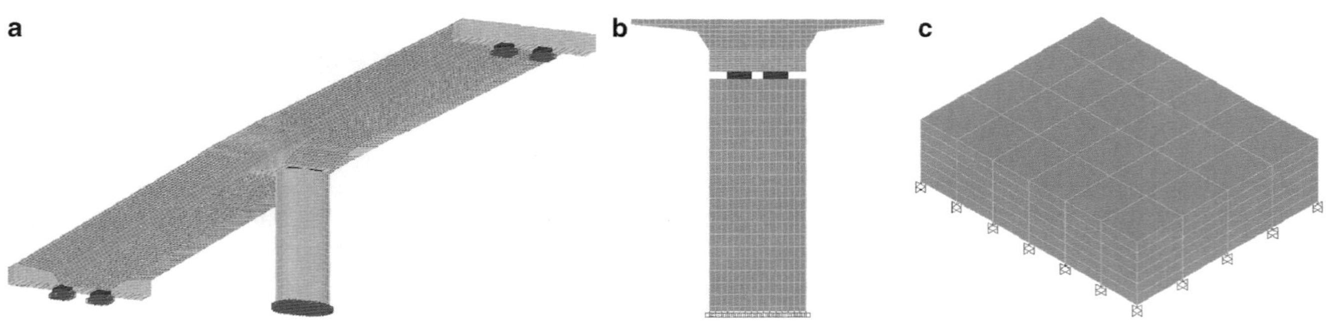

Fig. 1.6 Preliminary FE model: (**a**) FE mesh of deck; (**b**) modeling of transversal cross section; (**c**) detail of the isolator

Table 1.2 Correspondence between experimental (EMA) and analytical (FEA) mode shapes of the Preliminary FE Model. Natural frequency errors. $\Delta = 100 \times (p_{FEA} - p_{EMA})/p_{EMA}$; w.c. = without correspondence

EMA mode order r	EMA frequency p_r [Hz]	FEA mode order r	FEA frequency p_r [Hz]	Δ	MAC [%]
1	2.022 ± 0.001	1	2.062	1.99	99.4
2	3.053 ± 0.003	2	2.530	−17.13	92.8
3	3.180 ± 0.002	4	3.301	3.80	89.6
4	3.605 ± 0.002	3	3.246	−9.96	100.0
5	4.831 ± 0.011	5	3.862	−20.06	95.8
6	6.887 ± 0.046	7	7.129	3.51	98.0
7	6.934 ± 0.015	8	7.693	10.94	85.2
8	7.995 ± 0.005	10	10.259	28.31	89.7
9	9.107 ± 0.020	9	9.367	2.85	83.5
10	12.910 ± 0.025	w.c.	–	–	–
11	14.228 ± 0.081	16	19.585	37.75	53.3
12	14.433 ± 0.100	w.c.	–	–	–

bending mode of the bridge deck, can be explained by observing that this mode is weakly excited during testing, even along the transverse X-direction, because the position of the shaker was near a nodal region of this mode. To confirm the correspondence based on visual comparison, the Modal Assurance Criterion (MAC) matrix between experimental and analytical mode shapes has been evaluated [15]. MAC values collected in Table 1.2 have been evaluated with reference to the horizontal or vertical modal components for rigid-body and bending/torsional vibration modes, respectively. The correlation generally is extremely good, with MAC values higher than 0.85. Correlation becomes worse for experimental Mode 11 and EMA modes 10,12 have no analytical counterpart.

Comparing with the measured modal frequency values, it can be seen that the analytical FEM-based frequencies of the rigid-body vibration modes are appreciably lower than the corresponding experimental values for EMA Modes 2 and 5, with differences of about 17% and 20%, respectively. Experimental natural frequencies associated to bending/torsional vibration modes of the deck are overestimated, and errors generally are of order of few points per cent, with the exception of large discrepancies on higher order modes, namely about 28% and 38% on EMA Modes 8 and 11, respectively.

1.5 A Refined FE Model of the Bridge

An analysis of the differences in the numerical and experimental natural frequencies of the Preliminary FE Model implied that there was a problem with uncorrected modeling of the bridge. Moreover, the fact that experimental frequency values of rigid-body modes are underestimated and that natural frequencies of bending and torsional deck modes are overestimated, suggests that simultaneous underestimates and overestimates of inertia/stiffness properties may be present in the Preliminary FE model of the bridge. In addition, the large difference between transverse K_x and longitudinal K_y stiffness of the isolators (that, in principle, should be approximately equal) is hardly justified.

An extensive numerical/analytical investigation has been carried out in order to give an interpretation of the experimental results and, ultimately, to find an Refined FE model of the bridge. Referring to the paper [13] for a detailed description of the steps and results of the identification procedure, here the main improvements made on the Preliminary FE model and a discussion on their quantitative effects are summarized.

One of the improvements concerns the influence of the footways, which were neglected in previous stage. A numerical investigation performed on the basis of the Preliminary FE Model, in fact, showed that the footway mass, although marginal in comparison with the total mass of the bridge deck (about the 7%), cause no negligible reduction in natural frequencies, especially on torsional modes. It can be shown that this follows from the fact that the additional mass is located on regions of high sensitivity for those modes, or, equivalently, on regions with large modulus of modal displacement. The total mass of the bridge deck, footways included, is now $M' = 1{,}735$ t. Conversely, the stiffening contribution offered by the footways can be considered negligible, due to the presence of large longitudinal voids in the downstream side and because of the small size dimensions of the footway located on upstream side, see Fig. 1.1.

Another important point is connected with the influence of the in-plane transverse deformability of the deck and of the deformability of the pier on the estimate of the isolator stiffness. These two phenomena were neglected in previous stage and the assumption of rigid-body motion of the deck for fixed supports of the isolators was assumed. Actually, the analysis of the

Table 1.3 Correspondence between experimental (EMA) and analytical (FEA) mode shapes of the Refined FE Model. Natural frequency errors. $\Delta = 100 \times (p_{FEA}\text{-}p_{EMA})/p_{EMA}$; w.c. = without correspondence.

EMA mode order r	EMA frequency p_r [Hz]	FEA mode order r	FEA frequency p_r [Hz]	Δ	MAC [%]
1	2.022 ± 0.001	1	1.985	−1.84	99.4
2	3.053 ± 0.003	2	2.857	−6.41	87.3
3	3.180 ± 0.002	3	3.165	−0.48	89.5
4	3.605 ± 0.002	4	3.369	−6.56	100.0
5	4.831 ± 0.011	5	4.611	−4.56	94.3
6	6.887 ± 0.046	7	6.809	−1.13	97.5
7	6.934 ± 0.015	6	6.605	−4.75	83.6
8	7.995 ± 0.005	9	8.829	10.43	89.0
9	9.107 ± 0.020	10	8.903	−2.24	73.7
10	12.910 ± 0.025	w.c.	–	–	–
11	14.228 ± 0.081	15	16.488	15.97	56.5
12	14.433 ± 0.100	w.c.	–	–	–

deformed shape of the FEA transverse rigid-body mode shows a not negligible transverse deformability of the deck in the horizontal plane. Moreover, experimental FRF measurements show appreciable flexibility of the top section of the pier in the global dynamic behavior of the bridge, both for transverse and longitudinal vibrations. Therefore, it is reasonable to assume that this additional flexibility may affect the procedure used in previous section to estimate the shearing stiffness K_x, K_y of the isolators.

Taking into account the above considerations, the longitudinal stiffness of the isolators K_y was estimated by considering the influence of the compliance introduced by the pier and assuming rigid-body motion along the longitudinal direction of the bridge. The pier was described by a Timoshenko beam and, by imposing the EMA frequency value 3.61 Hz, it was found $K_y = 172.4$ MN/m. Concerning the transverse direction, the in-plane vibrations of the bridge were modeled by a flexible beam under three elastic supports, two at the abutments and one at the pier. By imposing the coincidence of the EMA frequency value 3.05 Hz it was found $K_x = 151.2$ MN/m.

The difference between the identified stiffness values for motions along the longitudinal and transverse direction is significantly reduced with respect to the previous estimate, and it is around 12%. This result corroborates the assumption of considering a unique value for both K_x and K_y, corresponding to the average value of about 161.8 MN/m. This optimal value is about 2.2 the nominal value, in agreement with the results found in [9]. Table 1.3 shows the correspondences between EMA and FEA vibration modes of the Refined FE Model of the bridge, and corresponding natural frequency values. It can be seen that there is a rearrangement of the modes with respect to the Preliminary FE Model, and that EMA Modes 6,7,8,9,11 now correspond to FEA Modes 7,6,9,10,15, respectively. Modeling errors are generally of order of few points per cent for the lower vibration modes. Underestimates are of order 4–6% for rigid-body motions, and errors on EMA Modes 8 and 11 are about 10 and 16%, respectively. We refer to the paper [13] for a further improvement of this model.

1.6 Conclusions

The results of a series of harmonic forced vibration tests on an isolated bridge have been presented and interpreted in this paper. An identification procedure based on modal analysis and finite element modeling has been adopted for the characterization of the bridge and, particularly, for the identification of the stiffness of the bearing devices supporting the deck superstructure. The research allowed to identify a baseline model of the bridge that will be useful for long-term maintenance plans and for possible future investigation of diagnostic character.

References

1. Aktan AE, Farhey DN, Helmicki AJ, Brown DL, Hunt VJ, Lee K-L, Levi A (1997) Structural identification for condition assessment: experimental arts. J Struct Eng ASCE 123(12):1674–1684
2. Benedettini F, Zulli D, Alaggio R (2009) Frequency-veering and mode hybridization in arch bridges. In: Proceedings of IMAC XXVII, Orlando

3. Beolchini GC, Vestroni F (1997) Experimental and analytical study of dynamic behavior of a bridge. J Struct Eng ASCE 123(11):1506–1511
4. Brownjohn JMW, Moyo P, Omenzetter P, Yong L (2003) Assessment of highway bridge upgrading by dynamic testing and finite-element model updating. J Bridge Eng ASCE 8(3):162–172
5. Conte JP, Xianfei H, Moaveni B, Masri SF, Caffrey JP, Wahbeh M, Tasbihgoo F, Whang DH, Elgamal A (2008) Dynamic testing of Alfred Zampa Memorial Bridge. J Struct Eng ASCE 134(6):1006–1015
6. Gentile C, Gallino N (2008) Condition assessment and dynamic system identification of a historic suspension foot-bridge. Struct Control Health Monit 15(3):369–388
7. Megalhaes F, Cunha A, Caetano E (2008) Dynamic monitoring of a long span arch bridge. Eng Struct 30(11):3034–3044
8. Chaudhary MTA, Abe M, Fujino Y (2002) Role of structural details in altering the expected seismic response of base-solated bridges. Mech Syst Signal Process 16(2–3):413–428
9. Ntotsios E, Karakostas C, Lekidis V, Panetsos P, Nikolaou I, Papadimitriou C, Salonikos T (2009) Structural identification of Egnatia Odos bridges based on ambient and earthquake induced vibrations. Bull Earthq Eng 7:485–501
10. Dilena M, Morassi A (2011) Dynamic testing of a damaged bridge. Mech Syst Signal Process 25(5):1485–1507
11. Dilena M, Morassi A, Perin M (2011) Dynamic identification of a reinforced concrete damaged bridge. Mech Syst Signal Process 25(8):2990–3009
12. Morassi A, Vestroni F (2008) In: Morassi A, Vestroni F (eds) Dynamic methods for damage detection in structures, vol 499, CISM International Centre for Mechanical Sciences. Springer, Wien
13. Bedon C, Morassi A (2012) Dynamic analysis of a seismically isolated bridge, Under review, 2012
14. Hamed E, Frostig Y (2006) Natural frequencies of bonded and unbounded prestressed beams-prestressed force effects. J Sound Vib 295(1–2):28–39
15. Ewins DJ (2000) Modal testing: theory, practice and application, 2nd edn. Research Studies Press, Baldock

Chapter 2
Cable-Stayed Footbridge: Investigation into Superstructure and Cable Dynamics

Reto Cantieni

Abstract A 35 year's old cable-stayed footbridge was investigated into analytically and experimentally to becoming able to rate it's state of health. The paper presents the dynamic tests performed on this twin-32-m-span prestressed concrete structure and the respective results. Besides the main girder the bridge exhibits an A-shaped pylon as well as eight cables consisting of posttensioned parallel wire tendons. On the one hand, the tests covered experimental modal analyses of the whole structure including main girder, pylon and cables as well as of two of the cables as isolated structures under ambient excitation conditions. In addition, the structure was loaded with a jogger crossing the bridge and the maximum cable response to manual excitation was determined. The latter is not discussed here. The bridge state of health could be rated "satisfying".

Keywords Ambient excitation • Dynamic behavior • Footbridge • Jogger excitation • Modal analysis

2.1 Introduction

Considering the fact of a cable-stayed footbridge cable having failed in the northern part of Switzerland, the owner of Oberwies Footbridge, the Swiss Federal Roads Office, FEDRO, decided to check the state of health of this 35 years old bridge (Fig. 2.1). The reason for being concerned is straightforward: Oberwies Footbridge is crossing Motorway A1, one of the main motorway arteries in Switzerland, in the Zürich neighborhood. A collapse of Oberwies Footbridge would close all eight lanes of this motorway. Already today, traffic jams occur in both directions on a regular basis, every morning and afternoon/evening.

Visual inspection of the bridge did not reveal any significant damage. This also included close-up inspection of the tendons in the anchoring points vicinity. In addition, analytically checking the bridge's load carrying capacity under static loads showed one minor weak point only (pylon cross girder carrying the main girder). However, visual inspection also revealed that the bridge is vibrating significantly when trucks pass underneath at a speed of between 60 km/h and the allowed maximum, 100 km/h. Also, passing joggers excited the structure to perceptible vibrations. It was therefore decided to perform an experimental investigation to becoming able to rate the bridge dynamic behavior.

2.2 The Bridge

Oberwies Footbridge consists of a cast in-situ and posttensioned main girder, a reinforced concrete pylon and eight cables (Fig. 2.2). The cables are of the BBRV parallel wire type with the fixed anchor being covered by a concrete slab at the main girder and the mobile anchor being located at the pylon. The long cables consist of 55 wires 7 mm, the short ones of 36 wires 7 mm. The cable free length is some 24 and 11 m respectively. The pylon and the two abutments are supported by two large piles each.

R. Cantieni (✉)
rci dynamics, Structural Dynamics Consultants, 8600 Duebendorf, Switzerland
e-mail: reto.cantieni@rcidynamics.ch

A. Cunha (ed.), *Topics in Dynamics of Bridges, Volume 3: Proceedings of the 31st IMAC, A Conference on Structural Dynamics, 2013*, Conference Proceedings of the Society for Experimental Mechanics Series 38,
DOI 10.1007/978-1-4614-6519-5_2, © The Society for Experimental Mechanics, Inc. 2013

Fig. 2.1 Oberwies footbridge

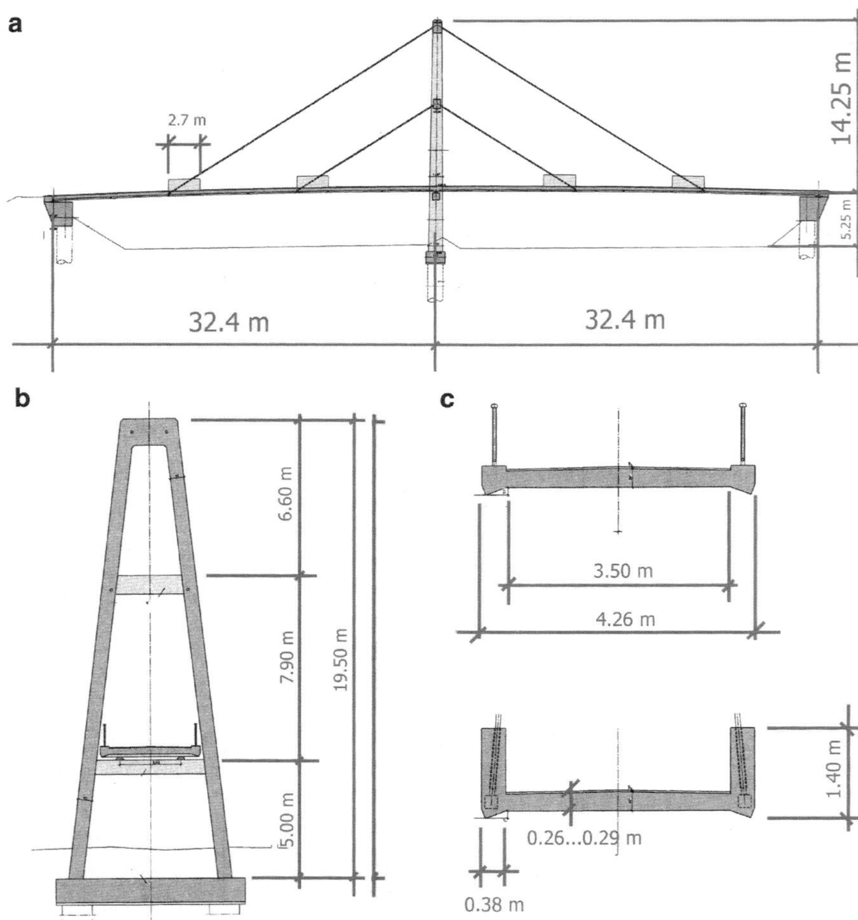

Fig. 2.2 Oberwies footbridge geometry

2.3 Experimental Modal Analysis of the Complete Structure Under Ambient Excitation

2.3.1 Strategy

The goal was to include main girder, pylon and cables in the modal analysis campaign. The main problem to be dealt with when defining the test strategy: How can we reach all desired measurement points without disturbing the heavy traffic? Furthermore, the final instrumentation layout should allow performing the modal test in 1 day using the 24-channel sensor/ frontend capacity available. One of the main parameters to consider here was: How long should we choose the time window per setup to be? To decide this, we need to know the bridge fundamental natural frequency. If we don't want to exclusively rely on an FE-analysis we have to perform a pilot test.

2.3.2 Pilot Test

To determine the bridge fundamental frequencies, one 3D- and one 1D-sensor (PCB 393B31, 10 V/g) were installed (Fig. 2.3). The sampling rate chosen was sR = 200 Hz, the time window length T = 1,800 s. The bridge fundamental natural frequency could be determined to f = 1.22 Hz (Fig. 2.4). It became also clear that the first four modes as identified in Fig. 2.4 were of the vertical bending type and No. 4–8 were of the torsion type respectively. Figure 2.5 indicates the shape of Mode No. 1 as derived from the two measurement point signals.

In addition, the opportunity was taken to optimize the instrumentation to be chosen to measure cable vibrations. As a result, PCB 393A03 (1 V/g) sensors were chosen because their low-frequency stability is better than the one of B&K 4508B (100 mV/g) sensors. The longer cable fundamental natural frequency was estimated to f = 2.9 Hz.

2.3.3 Logistics

To reach measurement points at the pylon and cables, a lifting device was necessary. The only place available for this was one of the motorway lanes adjacent to the pylon (Fig. 2.6). Stressing the Cantonal Highway Administration possibilities to the max left a time window for the lifting device being operational from 9.30 a.m. to 3.30 p.m.

Fig. 2.3 Pilot test: main girder instrumentation chosen

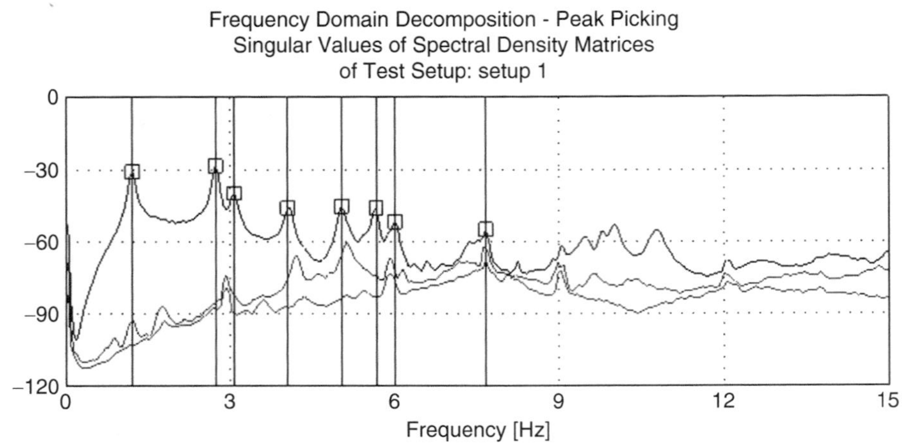

Fig. 2.4 Pilot test: FDD SVD diagram

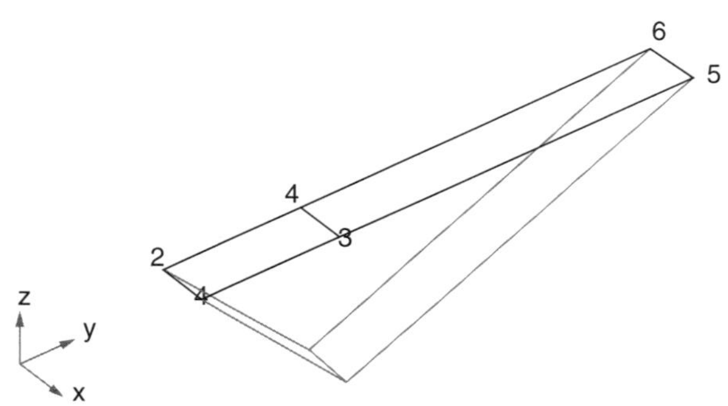

Fig. 2.5 Pilot test: shape of mode 1, f = 1.22 Hz

Fig. 2.6 How to reach pylon and cable measurement points

Fig. 2.7 Measurement sections. *Red*: main girder and pylon, *green/blue*: cables Zürich/Winterthur side

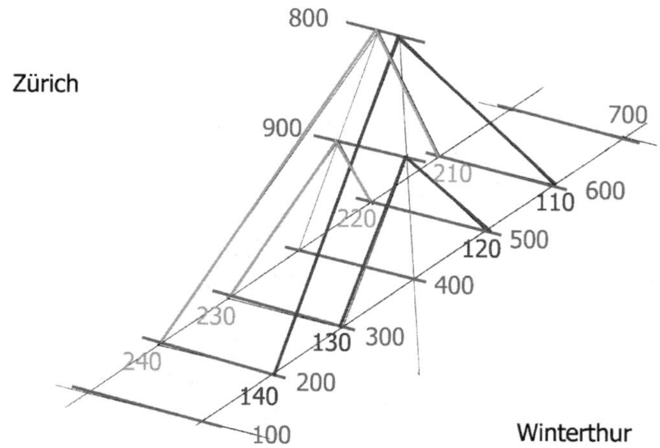

Fig. 2.8 Setup 1: *black*: references, *red*: rovers on main girder/pylon, *Green*: Rovers on cables

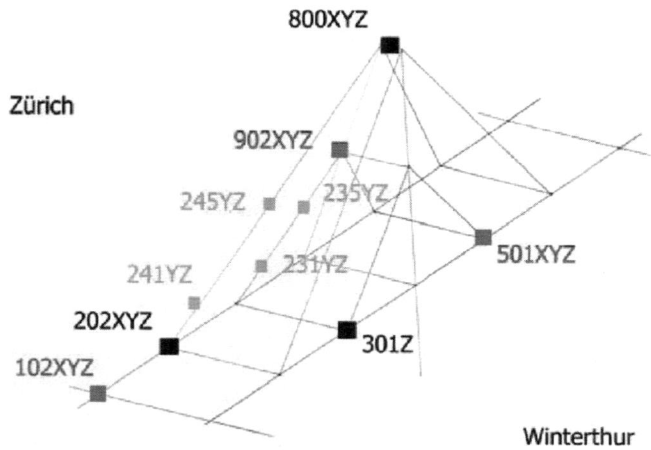

2.3.4 *Modal Analysis Test Instrumentation*

Considering the fundamental frequency being f = 1.22 Hz, the time window length per setup was chosen to 30 min. As a result of several optimization steps, it was decided to use five setups with 16 DOF's on the bridge girder and pylon and eight DOF's on one of the cables. Reference DOF's were located at the pylon tip (3D), at a main girder 0.33 L point Zürich side (3D) and at a main girder 0.66 L point Winterthur side (1D-vertical). Figure 2.7 presents the measurement section layout, Fig. 2.8 the instrumentation of Setup 1. The roving procedure from Setup 1 to Setup 5 is illustrated in Fig. 2.9.

This instrumentation schedule was possible through installing of the reference DOF's at the pylon tip, the rover at the pylon cross girder (902) and the points on the cables 240/230 the day before the tests only. The remaining installation: Measurement center, rovers at the concrete girder, cabling of everything, was performed the test day between 7 a.m. and 9.30 a.m. This allowed starting with Setup 1 the test day as early as possible: 09.30 a.m.

2.3.5 *Modal Analysis Test Performance*

Fortunately, everything could be performed as planned and the five setup signals were acquired August 25, 2011, 3.30 p.m. (Figs. 2.10 and 2.11).

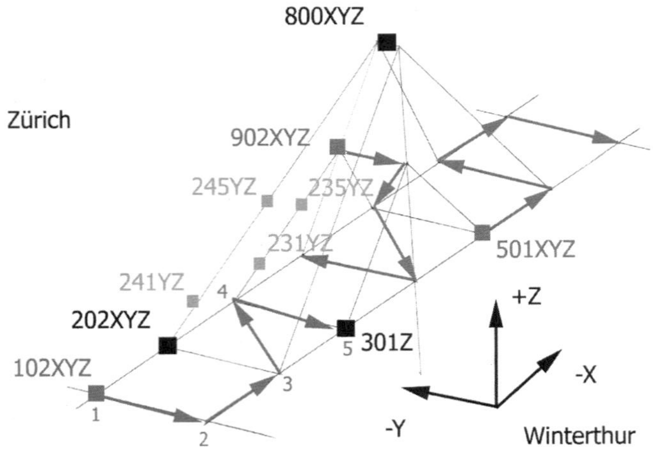

Fig. 2.9 Roving procedure. *Black*: references (2 × 3D, 1 × 1D). *Red*: Track of the rovers on main girder and pylon from Setup 1 to Setup 5 (3 × 3D). *Green*: The eight DOF's on cables 240/230, two on each cable, located at 2 m above bridge deck and at 0.4 L respectively, were moved to cables 140/130, 120/110 and 210/220 (see Fig. 2.7) for the setups 2–4. Setup 5 was performed without measurement points on the cables

Fig. 2.10 Sensors at the pylon tip (*left*) and at a cable (*right*)

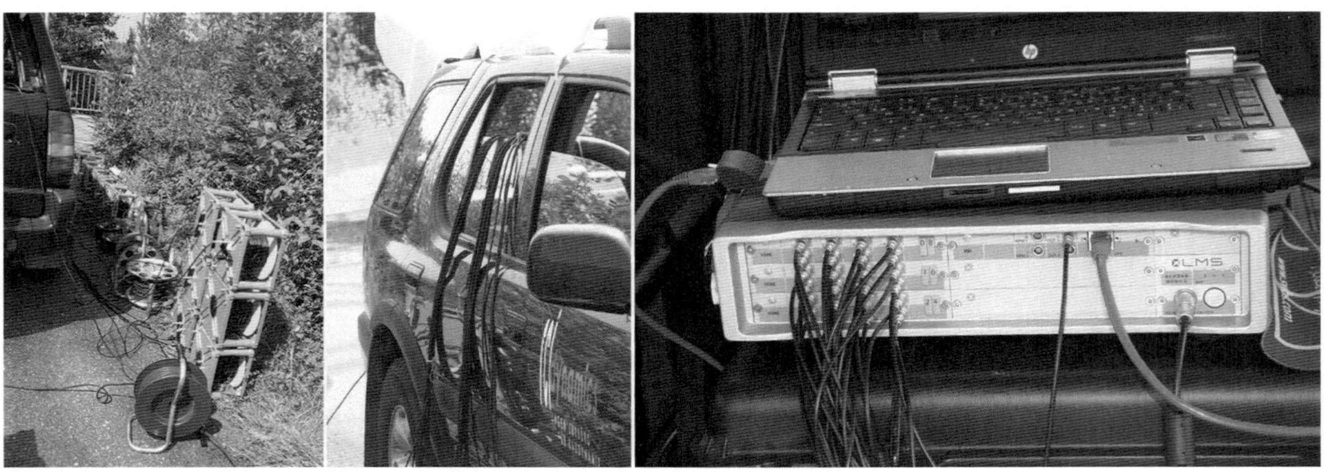

Fig. 2.11 Twenty four cable rolls (*left*), the measurement van (*middle*) and the frontend and laptop (*right*)

Fig. 2.12 EFDD SVD-diagram, raw version

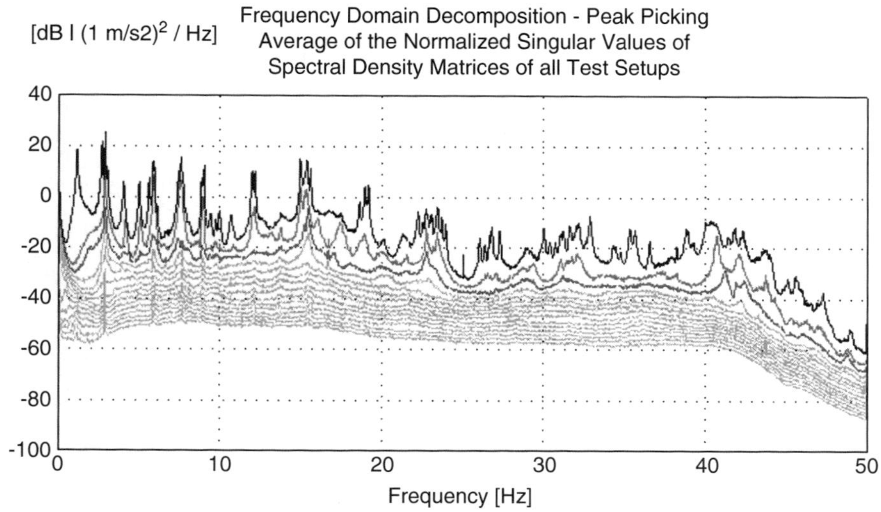

Fig. 2.13 EFDD SVD-diagram, f = 0…6 Hz. Bn = Bridge modes, SLn = Long Cables' modes

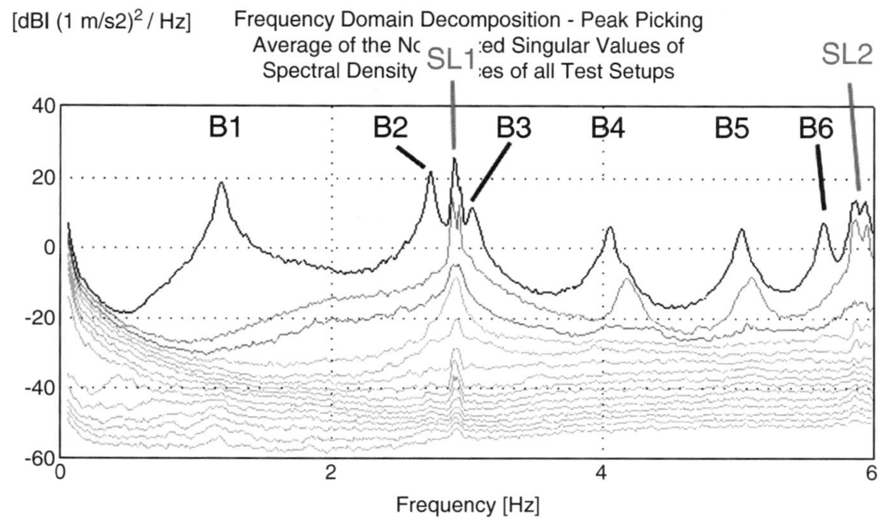

2.3.6 Modal Analysis Test Results

Data processing was performed using the Artemis Extractor EFDD routines. The results are presented in Figs. 2.12, 2.13, 2.14, and 2.15. SSI did not properly identify the low frequency modes. The raw data was decimated by a factor of 2. Reduction of the number of projection channels, e.g. to the number of the references used, seven, proved to be a bad idea. As no reference was located on a cable, this decreased the importance of the cable signals significantly and resulted in bad mode shapes in respect to the cable's shapes. The analysis frequency resolution was put to 4 K which resulted in a resolution $\Delta f = 0.012$ Hz. This allowed nice separation of the bridge and cable modes and also resulted in nice SVD diagram's shapes. However, the effort to optimize all parameters mentioned is definitely different from zero.

Checking the respective shape, it was easily possible to distinguish between bridge girder/pylon modes, modes of the long cables and modes of the short cables. Figure 2.12 gives the EFDD SVD-diagram raw version, Fig. 2.13 includes the results of the mode shape visual analysis for a zoomed-in frequency range. It is nice to see that the cable mode SVD lines pop up from the bottom lines indicating that (under ambient excitation) the bridge superstructure vibration is forced through cable vibrations at such frequencies.

Mode	Frequency [Hz]	Std. dev. Frequency [Hz]	Damping [%]	Std. dev. Damping [%]
B1	1.193	0.0073	1.89	0.14
B2	2.734	0.0029	0.63	0.062
B3	3.044	0.0083	0.88	0.13
B4	4.06	0.0053	0.74	0.20
B5	5.029	0.0065	0.58	0.079
B6	5.635	0.010	0.41	0.028
B7	9.433	0.035	0.53	0.18
B8	9.991	0.016	0.49	0.051
B9	10.72	0.015	0.74	0.10

Fig. 2.14 Nine bridge natural modes could be identified in the range f = 1.19...10.7 Hz. Damping of mode B1 and, to a lesser extent, of mode B3, is comparatively large. This is most probably due to the pylon modal motion being relatively large for these modes (see Chap. 7)

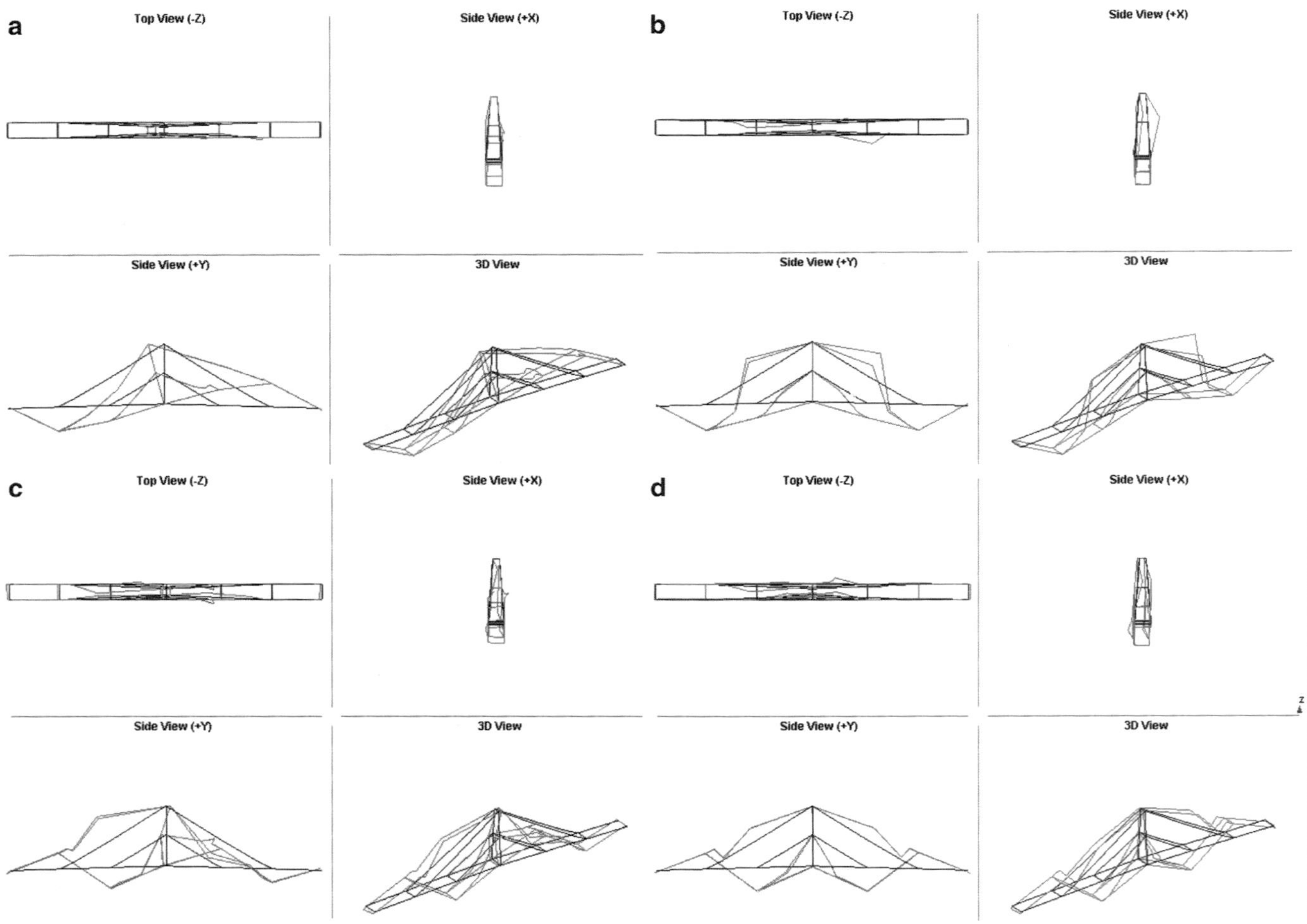

Fig. 2.15 (**a**) Mode 1, f = 1.19 Hz, $\zeta = 1.9$ %. (**b**) Mode 2, f = 2.73 Hz, $\zeta = 0.6$ %. (**c**) Mode 3, f = 3.04 Hz, $\zeta = 0.9$ %. (**d**) Mode 4, f = 4.06 Hz, $\zeta = 0.7$ %. (**e**) Mode 5, f = 5.03 Hz, $\zeta = 0.6$ %. (**f**) Mode 6, f = 5.64 Hz, $\zeta = 0.4$ %. (**g**) Mode 7, f = 9.43 Hz, $\zeta = 0.5$ %. (**h**) Mode 8, f = 9.99 Hz, $\zeta = 0.5$ %

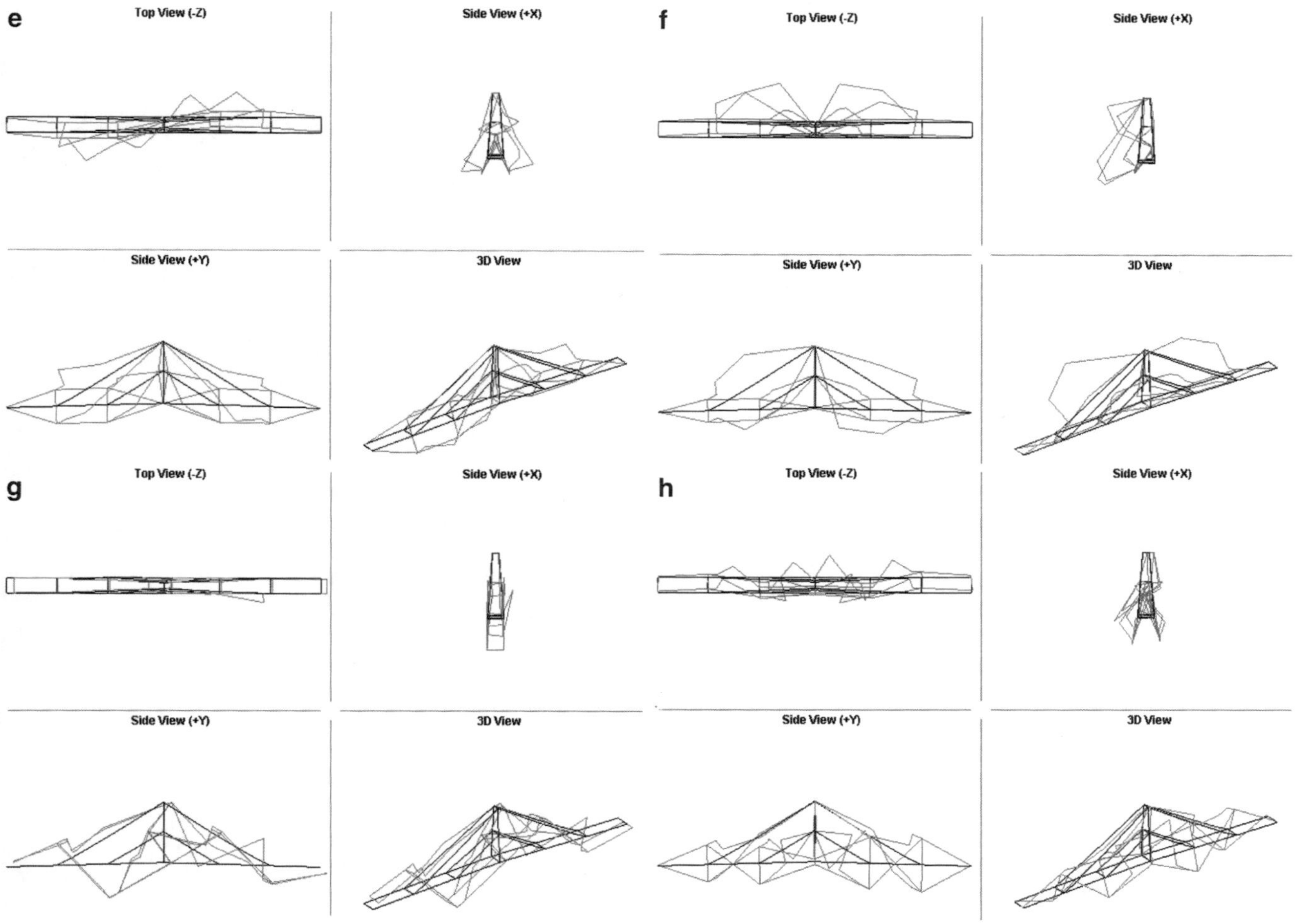

Fig. 2.15 (continued)

2.3.7 Modal Test Result Discussion, Bridge

The first mode B1 is dominated by a horizontal longitudinal pylon movement and an out-of-phase vertical motion of the two main girder spans. Its frequency f = 1.19 Hz is too low to be excited by walking people but it seems to be well suited to be excited by trucks passing underneath the bridge. This is quite straightforward. A truck of 20 m length travelling with 20 m/s (about 80 km/h) will produce an impulse with a length td = 1 s. In his famous book [1] from 1964, MIT-Professor John M. Biggs tells us that this is a very nice situation to produce a maximum system response of the Oberwies Footbridge exhibiting a fundamental period T = 0.84 s (Fig. 2.16).

Checking the bridge modes for susceptibilities versus pedestrian actions yields that, on the one hand, the bridge natural frequencies lie outside of the critical range for walking people (1.6…2.4 Hz) but, on the other hand, that modes B2 and B3 might be susceptible versus the action of joggers (2.5…3.5 Hz). As this had already become clear from the Pilot Test, jogger tests were planned to be included into the modal tests (see Chap. 5).

2.4 Experimental Modal Analysis of a Long and a Short Cable

Of course, it would have been possible to derive cable natural frequencies from the tests described in Chap. 3. In an attempt to determine the actual cable force through dynamic methods, special modal tests under ambient excitation were performed the day after the modal tests described in Chap. 3. The instrumentation is shown in Fig. 2.17. The sampling rate was again sR = 200 Hz, the time window length per setup started at 20 min and had to be reduced to 15 min due to the tight time schedule. The time pressure arose because the bottom point (No. 3 in Fig. 2.17) was measured in three different positions: at

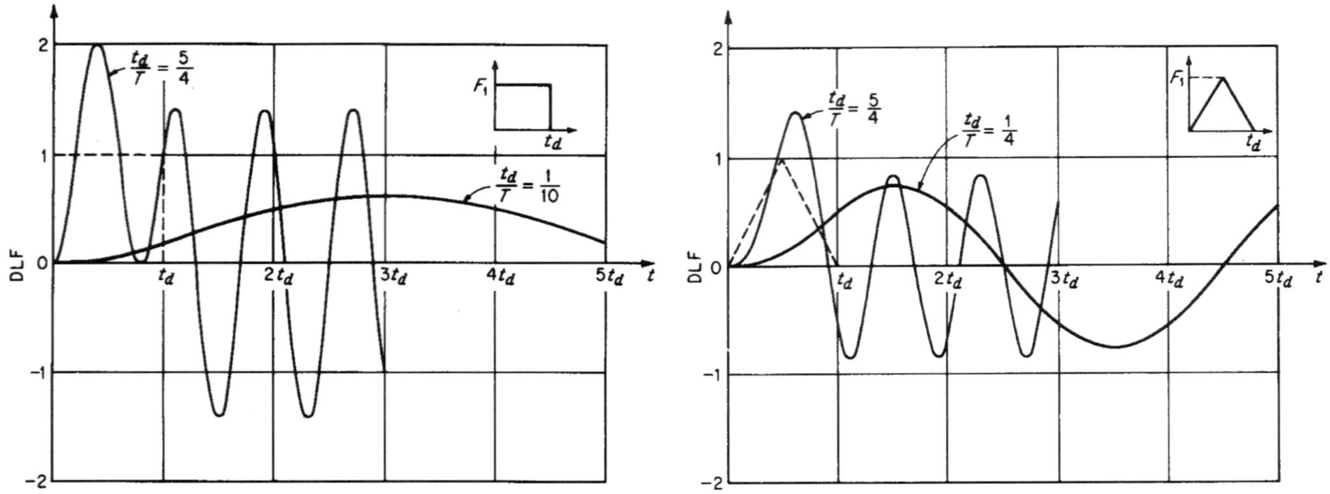

Fig. 2.16 Response of an undamped 1 DOF system with a period T to an impulse of length td (from [1])

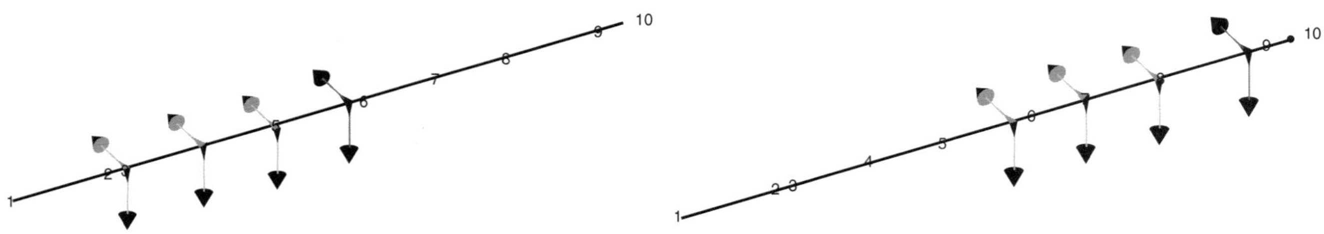

Fig. 2.17 Experimental Modal Analysis of the cables. *Blue*: References, *Green*. Rovers. Setups 1 (left) and 2

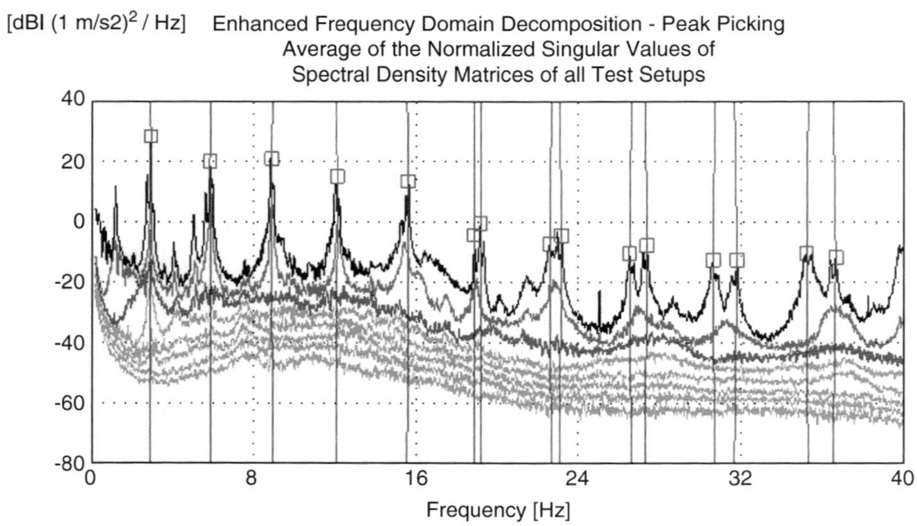

Fig. 2.18 EFDD SVD diagram for the long cable with the *bottom* point at 0.4 m

distances of 0.4, 0.8 and 1.2 m from the concrete slab end point. It was tried to identify the cable clamping conditions at the cable bottom end.

It can be seen from Fig. 2.18 that the cable modes between 0 and 40 Hz can easily be identified (of course, we know now the bridge modes) and that they all appear in an in-plane and an out-of-plane version. Data analysis was therefore performed for (X + Y), (X only) and (Y only). Due to space restrictions, this cannot be discussed here.

There is however a reason why we report the cable modal tests here (see below). Thus, for the sake of completeness, the numbers related to the first cable modes are given in Fig. 2.19 and some related mode shapes in Fig. 2.20.

Long Cable			Short Cable		
Mode	Frequency [Hz]	Damping [%]	Mode	Frequency [Hz]	Damping [%]
1	2.915	0.26	1	7.547	0.18
2	5.912	0.14	2	15.24	0.14
3	8.928	0.12	3	23.41	0.22
4	12.08	0.20	4	32.31	0.27
5	15.57	0.13	5	43.18	0.22
6	18.91	0.09	6	54.01	0.26

Fig. 2.19 First six long and short cable mode values. (X + Y) data processing

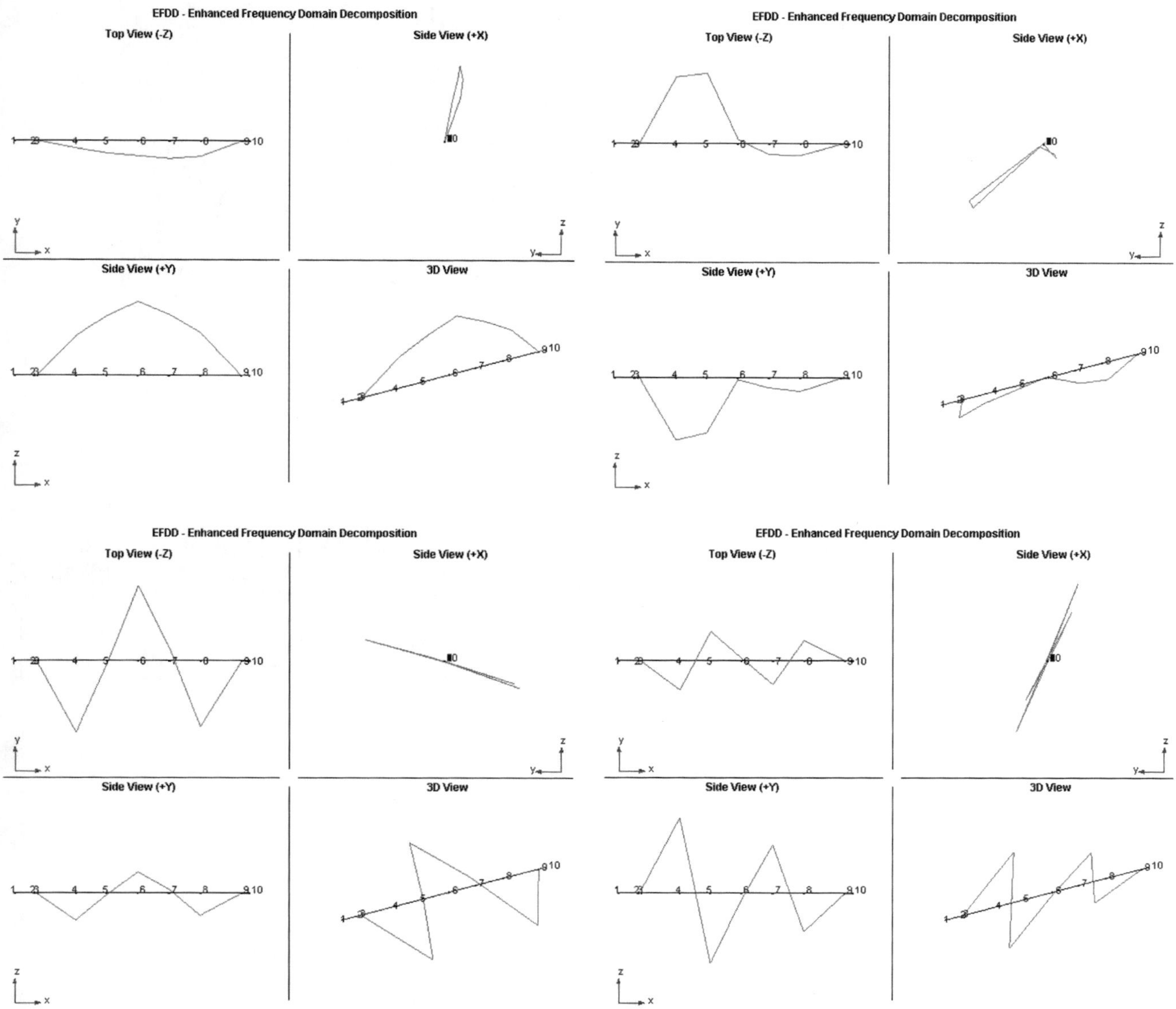

Fig. 2.20 Shapes of the long cable first four modes; (X + Y) data processing

2.5 Jogger Tests, Performance and Results

After each of the five Experimental Modal Test setups on the complete structure as described in Chap. 3, a series of jogger tests were performed. This was stretching the test time schedule to the max because the measurement chain sensitivity of 24 channels had to be adapted to the necessities and back. And the frontend software used in combination with the very slow Windows 7 operating system is not really up-to-par to cope with such a problem in an efficient way. However, we survived, loosing a lot of blood, sweat and tears (the outside temperature being some 30° Celsius). Figure 2.21 gives the jogger instrumentation as used to making the jogger keep a step pace of f = 2.73 Hz (B2) and to getting rid of Doppler Effects.

Firstly, there is one thing we can learn from the deflection signals (derived through double integration of the acceleration signals) and spectrum given in Fig. 2.22: The bridge does not like an input with f = 2.73 Hz, because this is not its

Fig. 2.21 Jogger test instrumentation. The frequency generator transforming the electronic sinus wave to an acoustic impulse signal is not visible

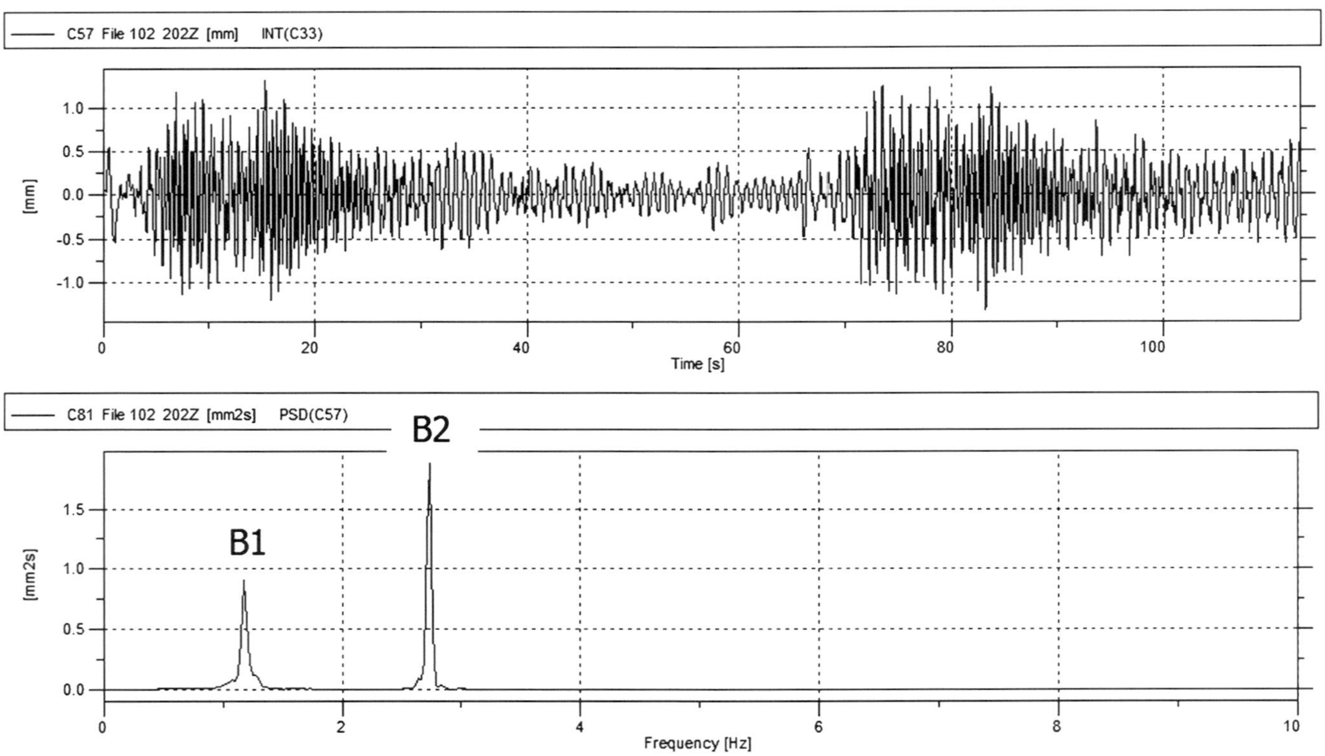

Fig. 2.22 Time signal and PSD-spectrum, point 202 (0.33 L), vertical. Jogger running over the bridge and back

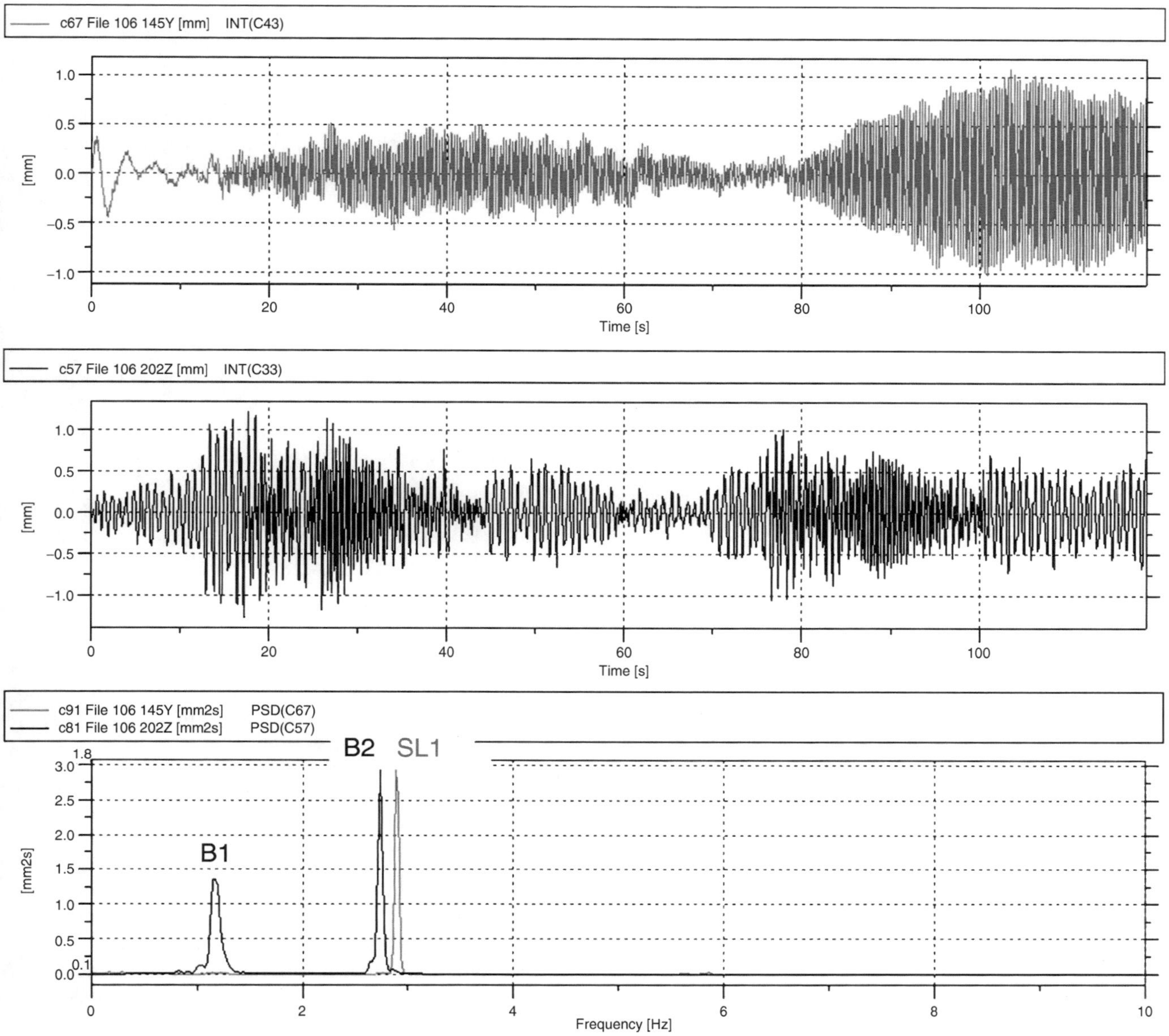

Fig. 2.23 The jogger crossing the bridge, and back, at a little too high pace. *Red*: Long cable out-of-plane displacement at 0.4 L, *Black*: Bridge vertical deflection at 0.33 L

fundamental frequency B1, f = 1.19 Hz. As long as the jogger is loading the bridge with f = 2.73 Hz, we can notice this in the vertical bridge response. As soon as the jogger stops, he was instructed to run over the bridge, wait for 30 s and run back, the bridge goes in the B1-mood, f = 1.19 Hz, it likes most. Just check the respective mode shapes to confirm this bridge-state-of-mood-philosophy.

Secondly, it was simply a consequence of being stubborn and also having a look at the cable vibration signals that the following fact was reckognized (Fig. 2.23): The jogger running at a little bit too high pace excited the long cable fundamental mode to a significant intensity. With a very slow decay after the jogger having left the bridge.

2.6 Cross Checks

Based on the experimental data, several checks were performed. These covered (stress, fatigue) behavior of main girder, pylon and cables under ambient and jogger loading conditions. We were not directly involved in the respective analytical procedures but we delivered data like relative cable motion (displacement). This was possible for the cases where the motion of the pylon fix-point, cable 0.4 L point and bridge deck at the lower cable fix point had simultaneously been measured only. No stress or fatigue problems could be identified.

2.7 Health Monitoring Using Dynamic Methods?

A commonly promoted procedure to monitor a structure's health is, a least for a level 1 damage detection, to monitor its natural frequencies. The fact of having "identified" Oberwies Footbridge twice, at the pilot and and the main tests, offers the opportunity to check the natural frequencie's stability versus temperature. The respective information is presented in Fig. 2.24. Temperature effects on natural bridge frequencies are significant for modes where the shape indicates significant connection of structural elements to soil. This is also discussed in [2].

2.8 Summary

Experimental Modal Analysis under ambient excitation is well suited for the identification of a 32-m-twin-span cable-stayed footbridge with a lot of highway traffic travelling underneath the bridge. Using 10 V/g sensors and choosing a long enough time window results in a very nice signal-to-noise-ratio. To measure the vibrations of stay cables with a 10–25 m length under ambient excitation, 1 V/g sensors are well suited.

Oberwies Footbridge exhibits a fundamental natural frequency B1, f = 1.19 Hz. This is not critical when it comes to pedestrian dynamic action. This mode is however well excited through trucks and trailers passing underneath the bridge with a speed v = 60...100 km/h. The headway is about 1 m which produces a nice air pressure wave during the vehicle passage. Cross-checks however revealed that the bridge response to such action may be clearly perceptible by humans but is of no danger to the bridge.

Oberwies Footbridge exhibits a second natural mode B2, f = 2.73 Hz. This is a frequency being nicely excited by joggers. However, respective tests showed that the bridge response to jogger excitation with f = 2.73 Hz is not critical because the respective frequency is not related to the first but to the second bridge mode. This means: Really critical states may occur if the exciting frequency corresponds to the structure's fundamental frequency and if the mode shape is similar to the static deflection shape forced by the exciter only.

Oberwies Footbridge long stay cables exhibit a fundamental natural frequency SL1, f = 2.9 Hz. This vibration is easily excited through a jogger's action. However, the excitation duration is too short to produce a "real" resonance problem.

Finally: The test proved that Oberwies Footbridge is a dynamically active structure without touching some critical limits. This may also be the reason for the bridge safely surviving 35 years without showing signs of distress.

Due to space restrictions we cannot discuss the attempts undertaken to deriving cable forces from dynamic measurements here. This is however a very interesting topic. Especially for cables where the clamping conditions are far from those applying to a string model.

Mode	Juli 20, 2011, 20 deg. C. Frequency f [Hz]	August 25, 2011 30 deg. C. Frequency f [Hz]	Δf [%]
1	1.22	1.19	2.5
2	2.73	2.73	0
3	3.08	3.04	1.5
4	4.05	4.06	0
5	5.03	5.03	0

Fig. 2.24 Oberwies Footbridge natural frequencies as a function of temperature

References

1. Biggs JM (1964) Introduction to structural dynamics. McGraw-Hill, New York/San Francisco/Toronto/London
2. Cantieni R (2012) Health monitoring of civil engineering structures – What we can learn from experience. In: Proceedings of 3rd IALCCE, international symposium on live-cycle civil engineering, Vienna, Oct 3–6 2012, p 400

Chapter 3
Analysis and Mitigation of Vibration of Steel Footbridge with Excessive Amplitudes

S. Pospíšil, S. Hračov, S. Urushadze, and D. Jermoljev

Abstract The article analyses the dynamics response of a footbridge crossing the Morava River between Slovak and Austrian border. After the construction, the bridge with very flexible construction showed excessive acceleration amplitudes in both vertical and horizontal direction when excited by usual pedestrians. Main footbridge part of 180 m length across waterway is a three span suspended steel structure. The longest middle span of 120 m length is hanged by four prestressed tendons over pylons, which are the part of prestressed lateral portals. Footbridge deck is made using triangular lattice girder with the top orthotropic plate. The bridge has a vertically curved shape. The adverse vibration has been successfully suppressed by multiple tuned dampers with the adjustable friction elements installed beneath the bridge deck. The experimental modal analysis comprising both forced vibration as well as ambient vibration from wind was carried out before and after the installation of vibration absorbers. The results of experimental tests were confronted with the outputs of numerical analysis.

Keywords Steel footbridge • Human induced vibration • Pedestrian comfort criteria • Tuned mass dampers

3.1 Introduction

Long-span footbridges rank among engineering structures with a high social significance. Their architecture design must in general satisfy matters such as attractive look, reliability and low weight. These requirements determine the bridge to be sensitive to the wind effects and the assessment against the wind becomes crucial. According to the character and the intensity of excitations, the adequate treatment to reduce or avoid the oscillations should be carried out also in cases when the vibration amplitudes could lead to the pedestrian discomfort or panic. In the terminal years of the last and the initial years of this century the number of footbridges has increased fast, as they provide shortcuts enabling pedestrians and cyclists to reach easily their destinations. While bridge structures are designed to withstand permanent load increased by a dynamic coefficient and not assessed with reference to user's comfort, the footbridge structure must resist both static and particularly dynamic loads. Moreover, its dynamic response must satisfy also the requirements of pedestrians' comfort usually expressed by response acceleration. From 1970 to the present the Central laboratory of the Institute of Theoretical and Applied Mechanics of the Academy of Sciences has tested some 15 footbridges of various type, either in situ or on their dynamic/aeroelastic models. The tests verified in particular the magnitude of dynamic response; the model tests provided also information on wind effect.

Regulations and standards, e.g. [1–4], recommend the avoidance of such structures, the vertical natural frequencies of which vary from between 1.6 and 2.4 to between 3.5 and 4.5 Hz (this second range is connected with the danger of second harmonic response excitation). However, these requirements cannot be fulfilled in all cases. The hatched zone in Fig. 3.1 is a zone through which the curves pass expressing the probable upper and lower limits of the vertical footbridge natural

S. Pospíšil (✉) • S. Hračov • S. Urushadze
ITAM AS CR, v.v.i., Prosecká 76, Prague 9 19000, Czech Republic
e-mail: pospisil@itam.cas.cz; hracov@itam.cas.cz; urushadze@itam.cas.cz

D. Jermoljev
EXCON a.s., Sokolovská 187/203, Prague 9 19000, Czech Republic
e-mail: jermoljev@excon.cz

A. Cunha (ed.), *Topics in Dynamics of Bridges, Volume 3: Proceedings of the 31st IMAC, A Conference on Structural Dynamics, 2013*, Conference Proceedings of the Society for Experimental Mechanics Series 38,
DOI 10.1007/978-1-4614-6519-5_3, © The Society for Experimental Mechanics, Inc. 2013

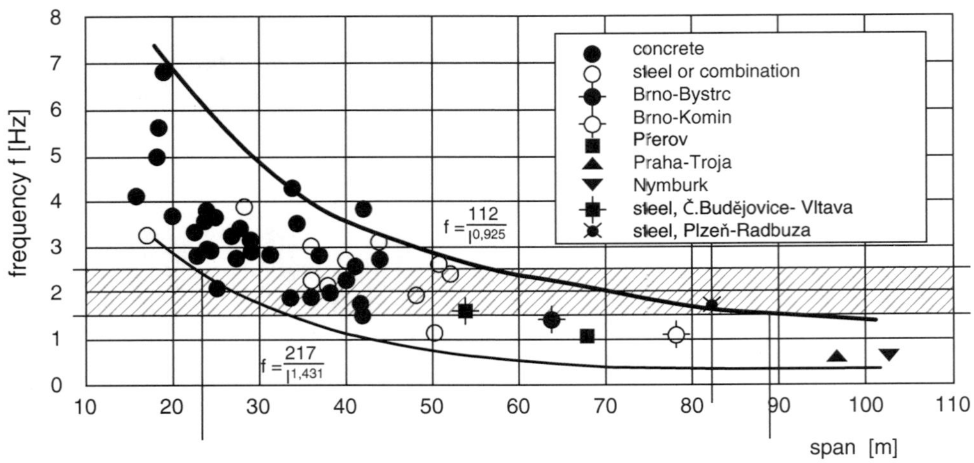

Fig. 3.1 The lowest vertical natural frequency versus main span of the footbridge

frequency. The points of intersection of these curves with the zone reveal that the footbridges with the span (of the main opening) between 23 and 89 m are unable to fulfill the aforementioned requirement. Most tested footbridges did lie within these span limits and, therefore, had to be subjected not only to theoretical, but also to experimental analysis.

The article analyses the dynamics response of a footbridge crossing the Morava River between Slovak and Austrian border. After the construction, the bridge with very flexible construction showed high acceleration amplitudes in both vertical and horizontal direction when excited by pedestrians. Tuned mass dampers were installed into the bridge deck for suppressing of excessive vibrations. The experimental modal analysis comprising both forced vibration as well as ambient vibration from wind was carried out before and after the installation of vibration absorbers. The numerical analysis was also performed and results were compared with output data from measurements of a real structure.

3.2 Bridge Description

Footbridge body is made using triangular lattice girder with the top orthotropic plate deck. The bridge has a vertically curved shape, see Fig. 3.2. Main footbridge part of 180 m length across waterway is a three span suspended steel structure. The longest middle span of 120 m length is hanged by four prestressed tendons over pylons, which are the part of prestressed lateral portals.

The axes of the bridge form an triangle cross-section with a vertical plane of symmetry and the lower vertex lying in that plane. Three main longitudinal tubes pass through the vertices of the triangle (upper tubes with dimensions 178 × 10 mm, lower tube with dimensions 356 × 12.5 (20) mm). The side walls form a triangular tubular diagonals and verticals (133 × 8, 133 × 10, 133 × 16). The upper deck wall forms a triangle with the crossbeams 2.5 m distant mutually. The crossbeams with the axial distance 2.5m support orthotropic deck plate reinforced by longitudinal ribs 100 × 10 with the distance of 400 mm. The main bridge deck has 120 m and it is supported with tree-like bearing. The bridge deck is supported by the Macalloy hangers (type 460 M100) at the quarters of the main field, two pairs of these rods are connected with outer fields. Portals are complemented by four rods forming the lateral support platform while increasing its lateral stiffness. Details about the construction can be seen in the reference [5].

3.3 Numerical Model, Load Model

For dynamic analysis a computational model of the footbridge has been created in the software SCIA Engineer 2009, which is designed for calculations of structures using a finite element method. The model was built in three-dimensional space using bar, beam and plane elements, see Fig. 3.3. Damping matrix was considered in all calculations of response as proportional to a combination of mass and stiffness matrix. The damping ratio was chosen by low value 0.5 % due to steel structure for all eigen-modes, which were taken into the consideration.

In this paragraph the calculated modal parameters of model i.e. eigen-frequencies and eigen-modes are given and depicted respectively. Table 3.1 showed the first ten relevant natural frequencies lying in the range 0–5 Hz. The eigen-

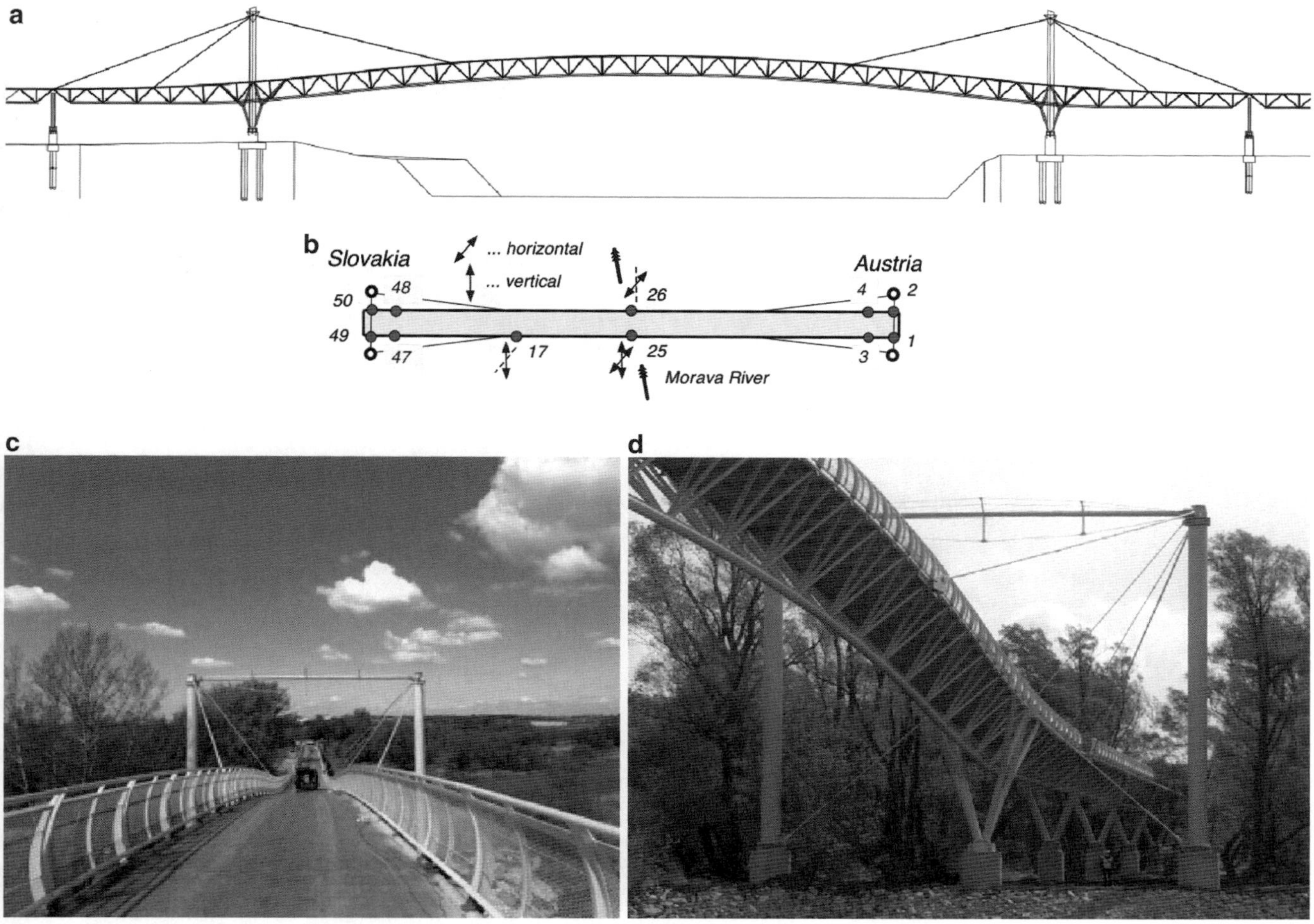

Fig. 3.2 (**a**) Bridge view; (**b**) Diagram of the bridge and the location of measurement points and orientation of sensors; (**c**,**d**) View on the bridge from the deck during measurements and from the left bank

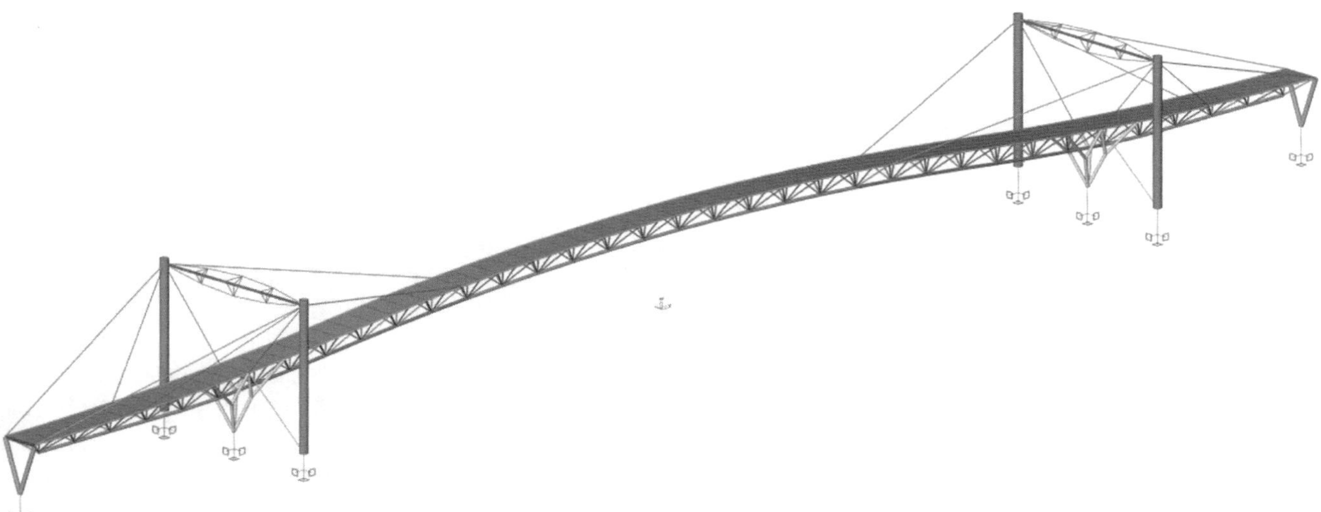

Fig. 3.3 Computational model of the footbridge

Table 3.1 Calculated eigen-frequencies and eigen-modes

Mode no.	f [Hz]	Character of eigen-mode
1	**1.10**	**Torsion + *Horizontal* bending**
2	**1.63**	**Vertical bending**
3	**2.15**	**Vertical bending**
4	**2.49**	**Vertical bending**
5	2.53	Torsion + *Horizontal* bending
6	3.12	Torsion + *Horizontal* bending
7	3.73	Torsion + *Horizontal* bending
8	**3.76**	**Vertical bending**
9	3.98	*Torsion* + Horizontal bending
10	**4.69**	**Vertical bending**

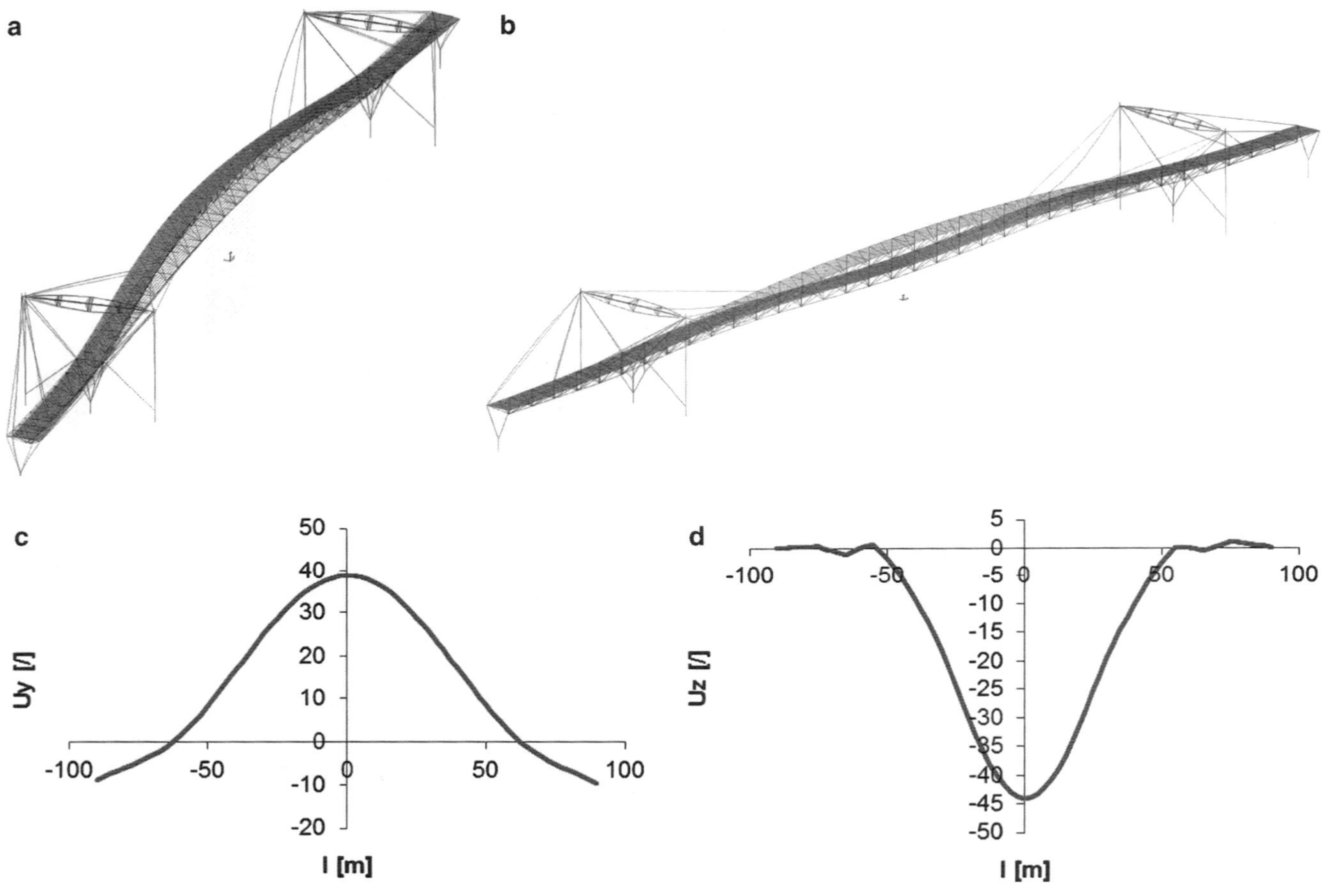

Fig. 3.4 First two eigen-modes of structure; (**a**) – axonometry and (**c**) – horizontal displacement of 1. lateral mode (f = 1.1 Hz); (**b**) – axonometry and (**d**) – vertical displacements of 1. Vertical mode (f = 1.63 Hz)

frequencies, which lie in the critical vertical (1–5 Hz) and horizontal (0.5–2.5 Hz) range of the pedestrian step frequencies are typed with bold. These eigen-frequencies and relevant mode shapes were analyzed with respect to the unacceptable vibration. Five modes shapes of the first ten are identified as a combination of torsion and bending, the prevailing direction is typed with italic. The first two eigen-modes are presented in Fig. 3.4 both in axonometry and in the plane view.

3.3.1 Comfort Criteria and Analysis of Human Induced Vibration

The effects of vibration on the human organism are usually expressed in terms of velocity as a function of frequency. This approach is suitable especially for higher frequency oscillation. For lower frequencies that are characteristic for flexible structures, the criterions are given in codes and hygienic regulations in the units of accelerations. The maximum accepted

Table 3.2 Comfort classes with acceleration ranges [6]

Comfort class	Degree of comfort	Vertical acceleration a [ms^{-2}]	Horizontal acceleration a [ms^{-2}]
1	Maximum	< 0.50	< 0.10
2	Medium	0.5–1	0.1–0.3
3	Minimum	1–2.5	0.3–0.8
4	Unacceptable discomfort	> 2.5	> 0.8

accelerations of bridge deck, which are recommended by Eurocode [4] are for vertical vibrations $a_{vert,max} \leqq 0.7\ ms^{-2}$ and for horizontal vibrations $a_{hor,max} \leqq 0.2\ ms^{-2}$ when normal traffic is considered or $a_{hor,max} \leqq 0.4\ ms^{-2}$ for the exceptional traffic cases. Comfort classes in terms of acceleration can be found in guidelines [2, 6] the latter are presented in Table 3.2.

According to the code [4], the comfort of pedestrians is not necessary to be assessed if the natural frequency of footbridge is higher than 5 Hz for vertical vibrations and 2.5 Hz for horizontal vibration respectively. According to [6] a critical range of natural frequency of bridges for possible excessive human induced oscillations is for vertical direction from 1.25 to 4.6 Hz and for horizontal direction 0.5–1.2 Hz. The natural frequencies of analyzed footbridge fall into these ranges and therefore the structure should be assessed from the point of view of the comfort criteria.

3.3.2 Model of the Pedestrian Induced Dynamic Load and Dynamic Analysis

The dynamic response of numerical model of footbridge was analyzed for load cases presented in guidelines [6]. Load cases are defined separately for horizontal and vertical direction and for individual natural frequency in a form of surface load:

$$p_{v,h}(t) = P \cdot \Psi \cdot \eta \cdot \sin(2\pi f_{v,h} \cdot t) \quad [Nm^{-2}] \tag{3.1}$$

where, P is the component of the force due to a single pedestrian with the step frequency $f_{v,h}$ which is assumed equal to the footbridge natural frequency (v. . .vertical direction: P = 280 N; h. . . horizontal direction: P = 35 N).

Ψ is the reduction coefficient taking into account the probability that the footfall frequency approaches the critical range of natural frequencies under consideration (0 ÷ 1).

η is the equivalent number of total number of pedestrians n on the loaded surface S:

$$\eta = 10.8\sqrt{\varsigma n}/S \quad [m^{-2}] ....for\ traffic\ density\ d < 1\ person/m^2 \tag{3.2}$$

$$\eta = 1.85\sqrt{n}/S \quad [m^{-2}] \ ....\ for\ traffic\ density\ d \geqq 1\ person/m^2 \tag{3.3}$$

ζ *is* the structural damping ratio

The load (3.1) respects different phase of individual point of bridge deck. To assess the bridge according to [4] and [6], three traffic density d were analyzed after consultation with a client:

(a) Weak traffic: d = 0.2 person/m^2 (14 kg/m^2)
(b) Dense traffic: d = 0.5 person/m^2 (35 kg/m^2)
(c) Very dense traffic: d = 1.0 person/m^2 (70 kg/m^2)

The calculated maximum acceleration of the steady state response of the bridge deck to load given by (3.1) are summarized for each natural frequency and traffic density in Table 3.3. Frequency f_{m+} represents the updated eigen-frequency of the structure that takes into account the added mass of pedestrians. Only resonant response for the first two vertical eigen-modes and first lateral eigen-mode were analyzed due to reduction coefficient Ψ equal or close to zero for all remaining modes. The above limits of acceleration with respect to code [4] are in Table 3.3 typed in bold. The limit values of acceleration in guidelines [6] are not given as strictly as those in [4]. The authors recommend an adapting of the degree of comfort to real traffic and requirements of the client.

The calculated vertical acceleration of resonant vibration in the first vertical mode was very high for all considered traffic densities. Thus it is the case, when pedestrians moving on the bridge would experience significant discomfort. In case of the second vertical eigen-mode, the limit [4] was exceeded only for very high traffic density. In case of resonant vibration in the first horizontal natural frequency the maximum acceleration are above limit for both dense and very dense traffic. However the lock-in effect should be taken into account for the lateral vibration. A significant synchronization between pedestrians

Table 3.3 Calculated maximum acceleration of resonant vibration of bridge deck in horizontal and vertical eigen-modes

Eigen-mode n.	f_0 [Hz]	d [person/m^2]	f_{m+} [Hz]	a_{max} [ms^{-2}]	a_{lim} [4] [ms^{-2}]	a_{lim} [6] [ms^{-2}]	Degree of comfort [6]
1 (hor)	1.1	0.2	1.07	0.136	< 0.2	0.1–0.3	Medium
1 (hor)	1.1	0.5	1.03	**0.217**	< 0.2	0.1–0.3	Medium
1 (hor)	1.1	1	0.98	**0.968**	< 0.2	> 0.8	Unacceptable
2 (ver)	1.63	0.2	1.59	**1.020**	< 0.7	1–2.5	Minimum
2 (ver)	1.63	0.5	1.54	**0.925**	< 0.7	0.5–1	Medium
2 (ver)	1.63	1	1.46	**1.861**	< 0.7	1–2.5	Minimum
3 (ver)	2.15	0.2	2.14	0.148	< 0.7	< 0.5	Maximal
3 (ver)	2.15	0.5	2.12	0.310	< 0.7	< 0.5	Maximal
3 (ver)	2.15	1	2.09	**0.990**	< 0.7	0.5–1	Medium

Table 3.4 Dynamic parameters of tuned mass dampers

Direction	Frequency f [Hz]	Gen. mass Meff [t]	Freq. f_{TMD} [Hz]	Mass m_{TMD} [kg]	Stiffness k_{TMD} [N/m]	Damping c_{TMD} [Ns/m]	Number [/]
Vertical	1.63	50.67	1.52	1,200	109,179	2,981	2
Horizontal	1.1	68.92	1.02	800	33,021	1,326	4

could occur and it could lead to an increase in amplitudes of vibration. In the document [6] the acceleration range of 0.1–0.15 ms^{-2} at which the appearance of the lock-in effect is given. This interval was achieved for all analyzed traffic cases.

More details about analysis of response can be found in report [7], which contains also the assessment of the footbridge on a more conservative load model suggested in article [8].

3.3.3 Design of Tuned Mass Dampers

The numerical analysis revealed that the footbridge is very sensitive to the excitation by the pedestrians. Therefore, damping should be increased by means of dynamic absorbers. In the present case, the design was complicated due to necessity of considering two "dangerous" modes to be damped out (the first vertical and the first horizontal mode). Moreover, the design dimensions, construction and the placement of the absorbers were limited by the space available under the bridge deck. Also environmental aspects of the design had to be reflected. Therefore, the robust tuned mass dampers with mechanical gadgets (springs and friction plates) were used to reduce the vibration associated with the first vertical and horizontal mode. The total weight of absorbers, their frequency and damping ratios were determined according to [9], see Table 3.4.

The total weight of absorbers is 1/22 of generalized mass of the structure vibration in respective mode shape, which meets the recommendation of optimal mass given in [10]. It has been calculated considering the case of weak traffic. The location of absorbers for vertical as well as for horizontal direction was proposed to be the most effective at half span, where both mode shapes have their maximal components. The layout and pictures of absorbers in the deck are presented in Fig. 3.5.

3.4 Measurement Description and Results

Measurements were carried out in two phases: before and after the installation of the tuned mass dampers. In the first phase, the frequencies and mode shapes were identified and compared to the calculation, see Table 3.5. To identify the eigen-frequencies (resonant frequencies), the bridge was excited by simple jumping. Also, the ambient vibration has been recorded on the bridge because of very strong wind that prevailed on the bridge during the measurement. We also conducted the measurement on the structure with an electrodynamic exciter placed on the construction in one-third of the span. Several cases of usual pedestrian traffic were finally modeled according to code [11] and the response was recorded, see Table 3.6. Vibration of the bridge was measured by the accelerometers (WILCOXON) that were placed on portable supports and moved along the bridge.

Figures 3.6a, b comprise the spectra of recorded horizontal and vertical vibration of bridge respectively which was loaded by the wind with mean wind speeds of about 10 m/s. The graphs of spectra give evidently the resonant frequencies, which

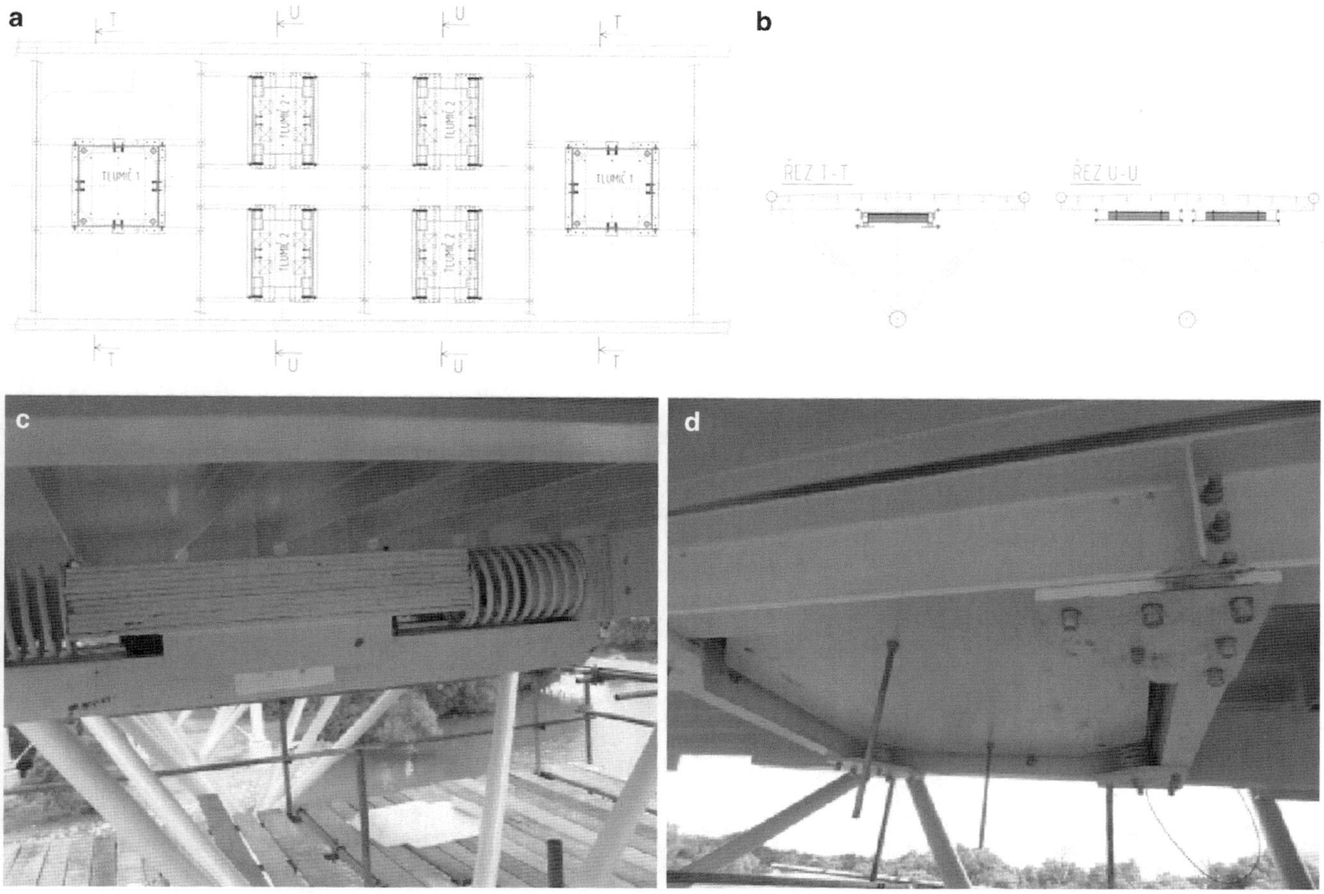

Fig. 3.5 (**a**,**b**) Part of the bridge drawing with tuned mass dampers; (**c**) tuned mass damper for horizontal direction; (**d**) tuned mass damper for vertical direction

Table 3.5 Values of the calculated and measured frequencies

Shape	f – numerics [Hz]	Mode shape	f – experiment I [Hz]	f – experiment II [Hz]
1	1.10	Torsion and horizontal bending	1.17	1.05
2	1.63	Vertical bending	1.72	1.53

Table 3.6 Values of the acceleration of the response

Load description	Limits [4] a [m/s^2]	Measurement I a [m/s^2]	Measurement II a [m/s^2]
Wind horizontal	0.2	0.08	0.01
Wind vertical	0.7	0.20	0.03
Two pedestrians crossing – horizontal	0.2	0.03	0.04
Two pedestrians crossing – vertical	0.7	0.50	**0.05**
Four pedestrians crossing – horizontal	0.2	0.08	0.05
Four pedestrians crossing – vertical	**0.7**	**0.95**	**0.08**
Crossing 2 + 2 pedestrians – horizontal	0.2	0.06	–
Crossing 2 + 2 pedestrians – vertical	**0.7**	**1.81**	**0.10**
Group crossing (seven persons) – horizontal	**0.2**	**0.16**	**0.06**
Group crossing (seven persons) – vertical	**0.7**	**2.96**	**0.10**

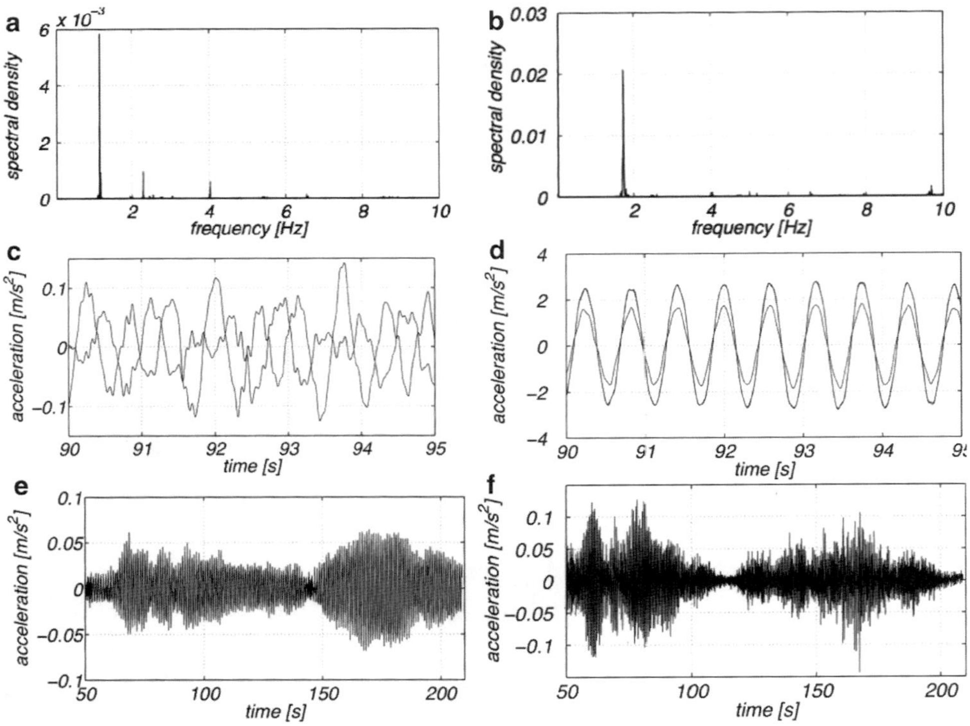

Fig. 3.6 (**a**) Spectral density of the horizontal response of mid-span of deck to the wind (without absorbers) (The first horizontal eigen-frequency is apparent – $f_1 = 1.17$ Hz). (**b**) Spectral density of the vertical response of mid-span of deck to the wind (without absorbers) (The first vertical eigen-frequency is identified – $f_1 = 1.72$ Hz). (**c**) Typical horizontal acceleration time histories of bridge deck (1/2 of span (point 25, 26 – see Fig. 3.2)) during a crossing by group of 7 people (without absorbers). (**d**) Typical vertical acceleration time history of bridge deck [1/2 (point 25) and 1/3 (point 17) of span] during a crossing by group of 7 people (without absorbers). (**e**) Typical horizontal acceleration time history of mid-span during a crossing by group of 7 people (with absorbers). (**f**) Typical vertical acceleration time history of mid-span during a crossing by group of 7 people (with absorbers)

can be compared with the calculation. The bridge was oscillating in the horizontal direction predominantly in resonance frequency $f = 1.05$ Hz with a shape corresponding to the first horizontal eigen-mode. Also in vertical direction one dominant resonant frequency peak $f = 1.53$ Hz was observed. The bridge vibrated in this case similar to a shape of the first vertical natural mode.

The slight decrease in frequency when compared to the first measurement was naturally due to the added mass dampers (5.6 t total), their supporting structures and temporarily mounted scaffolding. Mass of the bridge also increased due to mass of railing spanning in eight fields (20 m), which was not mounted during the first measurements. The overall increase in the weight of the bridge was 9.1 t including the absorbers. The graphs in Fig. 3.6c, d, e, f show selected examples of recorded excessive response for the same load case before and after insertion of absorbers respectively. In the Table 3.6, typical results are summarized, represented by the maximal value of the acceleration stationary response. It has to be noticed, that the response was very strong, especially for two following cases:

(a) The crossing of two person followed by another two pedestrians with the distance about 10 m having typical frequency around 1.2 Hz. In this case, a lock-in effect has been observed, when the pedestrians adapted their step frequency to match the frequency of the bridge.
(b) The group of 7 pedestrians, also with the usual step frequency, see Fig. 3.6c, d.

After installation of the TMDs, the response acceleration was suppressed to the accepted level, see e.g. Fig. 3.6e, f and Table 3.6.

3.5 Conclusions

The article presented a case study which deals with the dynamics response of a steel footbridge of a vertically curved shape and with a span 120 m. The bridge had very flexible construction and showed very high acceleration amplitudes in both vertical and horizontal direction when excited by usual pedestrians or even with a strong wind. In some circumstances – particularly when the bridge was crossed by two pairs of pedestrians with the distance between them approximately 10 m and by the group of people the accelerations reached almost triple of limit and acceptable values recommended by Eurocode [4]. The adverse vibrations have been successfully suppressed by the installation of multiple tuned mass dampers with the adjustable friction elements and with the low maintenance cost. They were mounted beneath the bridge deck. The experimental modal analysis comprising both the forced vibration as well as the ambient vibration from wind was carried out before and after the installation of vibration absorbers and compared with the numerical analysis. The footbridge is nowadays in permanent use for pedestrians and cyclists.

Acknowledgements The support was provided by the projects of the Ministry of Industry and Trade No. MPO TIP FR-TI3/654, Czech Grant Agency No. GAČR 103/09/0094 and research project RVO 68378297.

References

1. SWISS NORM SIA 160, Effects of Loads on Structures, (1989)
2. Comite Euro-Iinternational du beton, CEB-FIP Model Code 90p, CEB Bulletin D´information No. 213/214, (1993)
3. AASHTO – American Association of State Highway and Transportation Officials, Guide Specifications for Design of Pedestrian Bridges, (1997)
4. ČSN EN 1990 ed.2 Eurocode: Basis of structural design: Annex 2
5. Agócs Z, Vanko M (2012) Cyklomost Devínská Nová Ves – Schlosshof, Časopis Stavebnictví (in Slovak)
6. HIVOSS – Human Induced Vibrations of Steel Structures, Design of Footbridges, Guideline (2008) (http://www.stb.rwth-aachen.de/projekte/2007/HIVOSS/download.php) RFS2-CT-2007-00033
7. Hračov S, Pospíšil S (2011) Dynamické posouzení lávky pro pěší a cyklisty přes řeku Moravu z hlediska přijatelnosti vibrací od zatížení chodci (In Czech), Excon a.s.
8. Studničková M (2001) Evaluation of the footbridges according to European codes (in Czech), Stavební obzor No. 10/2001, p 296–298, Praha
9. Den Hartog JP (1956) Mechanical vibrations, 4th edn. McGraw-Hill, New York
10. Bachmann H et al (1995) Vibration problems in structures. Birkhäuser Verlag, Basel
11. ČSN 73 6209, Loading test on bridges (In Czech), Czech normalization institute, Prague, (1996)

Chapter 4
Change in Mass and Damping on Vertically Vibrating Footbridges Due to Pedestrians

Christos T. Georgakis and Nina G. Jørgensen

Abstract Pedestrian-induced footbridge vibrations are an issue that bridge designers often have to contend with. A plethora of research in recent years has led to the development of load models and procedures that allow for the determination of footbridge response. Nonetheless, measured footbridge responses often deviate from those predicted. One of the main deficiencies of the existing models and guidelines is the exclusion of the effect of changes in the footbridge's dynamic properties due to the presence of pedestrians. More specifically, any change in mass and/or damping that a pedestrian might introduce to a bridge will affect the bridges overall dynamic response. This effect is an element of what is often referred to as human-structure interaction. In this paper, the results of an experimental study to determine the change in mass and damping of a vertically vibrating footbridge due to traversing pedestrians are presented.

Keywords Vertical footbridge vibrations • Mass • Damping

4.1 Introduction

The majority of footbridges designed during the twentieth century were designed in a purely utilitarian manner, with the goal being the creation of a path for pedestrians to traverse between two points, without the impediment of e.g. vehicle traffic or bodies of water. Towards the end of the twentieth century this changed, as footbridge construction began to be viewed by architects and city planners as a tool for urban regeneration and as a means to create landmark structures and large-scale "sculptures". This shift in design philosophy, together with a reassessment our energy-intensive transportation infrastructure, has lead to a change in the perceived importance of footbridges. As a consequence, designers and engineers are paying greater attention to the aesthetic details and the servicability of footbridges, making sure that they are both beautiful and "comfortable".

Although footbridge comfort is often difficult to define, vibration comfort criteria within international codes and guidelines do exist [1, 2]. More often than not, the same codes and guidelines offer a means to assess the level of footbridge vibrations. Nevertheless, even though these codes and guidelines have often existed and been used, in some cases, for decades, it is not until recently that researchers and engineers have produced load models and procedures that allow for a more reasoned determination of footbridge response [3]. Still, even with the recent advances in load models and response prediction, measured footbridge responses often deviate from those predicted.

One of the main deficiencies of the new generation of models and guidelines is the exclusion of the effect of changes in the footbridge's dynamic properties due to the presence of pedestrians. Any change in mass and/or damping that a pedestrian might introduce to a footbridge will affect the footbridges overall dynamic response. This effect is an element of what is often referred to as human-structure interaction.

To the authors' knowledge, there has only been one other reported systematic attempt to determine the change of a footbridge's dynamic properties due to the presence of walking humans. Zivanovic et al. [4] showed an increase in damping of a 11.3 m long concrete footbridge, as a function of the number of traversing pedestrians. Unfortunately, the study was unsuccessful in generalizing the results in manner that would allow for their use on any footbridge or similar structure. Furthermore, no attempt was made to quantify a potential change in equivalent structural mass, due to the presence of pedestrians.

C.T. Georgakis (✉) • N.G. Jørgensen
Department of Civil Engineering, Technical University of Denmark, Building 118, Brovej, DK-2840 Kgs. Lyngby, Denmark
e-mail: cg@byg.dtu.dk; ninga@byg.dtu.dk

A. Cunha (ed.), *Topics in Dynamics of Bridges, Volume 3: Proceedings of the 31st IMAC, A Conference on Structural Dynamics, 2013*, Conference Proceedings of the Society for Experimental Mechanics Series 38,
DOI 10.1007/978-1-4614-6519-5_4, © The Society for Experimental Mechanics, Inc. 2013

In this paper, the authors present the results from an experimental campaign, aimed at determining the changes in mass and damping of a vertically vibrating footbridge due to pedestrians. The results are presented in a generalized manner, i.e. as the added/subtracted mass or concentrated damping per single pedestrian, at varying amplitudes of vibration and for varying probabilities of occurrence.

4.2 Methodology

The methodology used for the determination of the changes in mass and damping of a footbridge due to pedestrians is conceptually simple. First, a detailed experimental modal analysis of the laboratory footbridge without pedestrians is undertaken. From this, eigenmodes, frequencies, modal mass and damping of the empty footbridge are determined. Next, a hydraulic actuator is attached to the center of the footbridge; the purpose of this being the displacement- and frequency-controlled excitation of the footbridge at mid-span. The dynamic properties of the footbridge are once again determined through vertical dynamic actuation of the footbridge at mid-span. An excitation frequency sweep around the first natural frequency of the empty footbridge reveals the frequency at which minimum effort is needed for the actuator to excite the footbridge vertically. Once it is established that the footbridge's dynamic properties have not changed due to the introduction of the actuator, streaming pedestrians with varying flow rates are introduced to the footbridge. Longer measurements of the pedestrian-induced loads ensure that the excitation is stationary in nature. For the experiments reported herewith, vertical excitation was undertaken at predetermined bridge mid-span amplitudes of 1, 5 and 10 mm. The observed shift in frequency can be attributed to two separate effects, a change in structural mass and/or a change in damping.

If the vibration of the footbridge is restricted to motion in the first eigenmode, then the damped circular frequency of the empty footbridge will be:

$$\omega_D = \sqrt{\frac{k(1-\xi^2)}{m}} \tag{4.1}$$

where k is the modal stiffness, m is the modal mass and ξ is the modal damping ratio of the footbridge's first eigenmode. When pedestrians are traversing over the footbridge, a change in the footbridge's dynamic properties leads to a new damped circular frequency:

$$\omega'_D = \sqrt{\frac{k'(1-\xi'^2)}{m'}} \tag{4.2}$$

where k' is the modified modal stiffness, m' is the modified modal mass and ξ' is the modified modal damping ratio of the footbridge's first eigenmode. As the footbridge's modal stiffness is expected to remain constant, it is assumed that $k' = k$. By solving (4.2) for k' and substituting into (4.1), the footbridge's new modal mass as a result of the traversing pedestrians can be expressed as:

$$m' = m\frac{\omega_D^2(1-\xi'^2)}{\omega'^2_D(1-\xi^2)} \tag{4.3}$$

The footbridge's modified damped circular frequency is readily determined by performing an excitation frequency sweep, at predetermined displacement levels, close to the original circular frequency of the unloaded footbridge. The frequency at which the work required by the actuator is minimized is the modified natural frequency for the excited eigenmode.

The modified damping can be determined indirectly by computing the energy dissipated per vibration cycle by the footbridge. The energy dissipated per cycle by the footbridge is a function of the load imparted by the actuator and the velocity of the footbridge at the point of load application, so that [5]:

$$E_D = \int_0^T F(t)\nu(t)dt \tag{4.4}$$

where $F(t)$ is the dynamic load from the hydraulic actuator, $v(t)$ it the vertical velocity of the footbridge at mid-span, and T is the period of one cycle. The modified footbridge damping, expressed as an equivalent viscous damping ratio as a percentage of critical, is then found as [5]:

$$\xi' = \frac{E_D}{2\pi m' {\omega'_D}^2 X^2} \tag{4.5}$$

Where X is the maximum mid-span displacement of the footbridge. It can be seen from (4.5) that the equivalent viscous damping ratio is a function of the modified mass and, as such, the modified modal mass of the footbridge cannot be determined without employing an iterative approach.

4.2.1 Pedestrian Mass and Damping Coefficient

To generalize the findings, the individual contributions of each pedestrian to any potential change in footbridge mass and damping are determined. The change in the footbridge's damping ratio attributable to the pedestrians will be:

$$\xi_A = \xi' - \xi \tag{4.6}$$

The cumulative damping coefficient for all pedestrians can be written as:

$$C_A = 2m'\xi_A \omega'_D \tag{4.7}$$

Assuming a uniform distribution of the pedestrian population on the footbridge, c_A can also be written as:

$$c_A = \int_0^L c_p N \phi(y)^2 dy \tag{4.8}$$

where c_p is the damping coefficient of the individual pedestrian, N is the number of pedestrians on the footbridge, and $\phi(y)$ is the mode shape of the eigenmode under examination. From (4.7) to (4.8), the damping coefficient of the individual pedestrian is found as:

$$c_p = \frac{2m'\xi_A \omega'_D}{N \int_0^L \phi(y)^2 dy} \tag{4.9}$$

Similarly, it can be found that the mass contribution of the individual pedestrian is:

$$m_p = \frac{m' - m}{N \int_0^L \phi(y)^2 dy} \tag{4.10}$$

4.2.2 Experimental Setup

A simply-supported 16 m-long steel double U-beam footbridge, located in the Structures Laboratory of the Department of Civil Engineering at DTU (Figs. 4.1 and 4.2) was used as the basis structure for the experiments. The longitudinal beam profiles were UNP 350, with UNP 200 crossbeams placed at 1,400 mm intervals. For the experiments described herewith, masses were added to the footbridge crossbeams and to the center of the footbridge to increase the modal mass and thus decrease the footbridge's frequency to a level close to the expected mean pacing frequency of the pedestrians. The total mass of the footbridge was 5,224 kg.

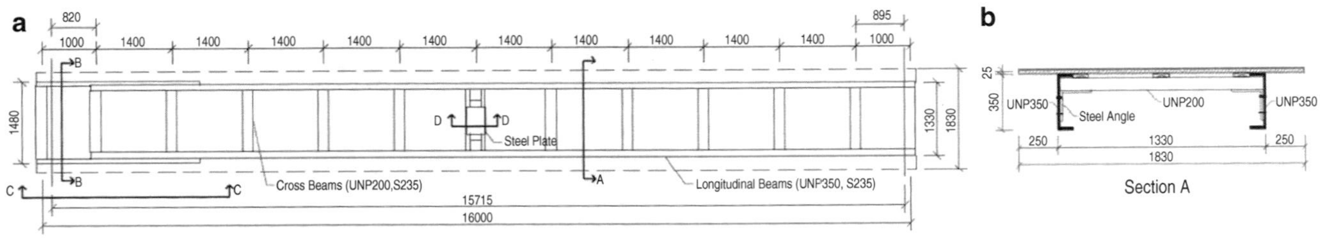

Fig. 4.1 (**a**) Elevation and (**b**) cross-section of the laboratory footbridge structure

Fig. 4.2 Pedestrians traversing the 16 m long DTU laboratory footbridge

The platform on which the pedestrians walked was made of medium density fiberboard (MDF) plates with a width of 1,840 mm. The plates were attached to the footbridge steel frame with screws. Wooden platforms of approximately 1.5 × 1.5 m^2 were constructed on either side of the footbridge to act as pedestrian launching platforms and turn-around areas. Stairs were made leading up to the platforms. The footbridge walking level was approximately 1.2 m above the laboratory floor. There was no hand railing on the footbridge.

Actuation of the footbridge during the tests was provided by a 100 kN Instron 8,500 hydraulic actuator, placed vertically beneath the center of the bridge (see Fig. 4.3). A calibrated Instron 8,516 load cell was used to measure the force between the actuator and the footbridge throughout the tests.

4.3 Experiments

4.3.1 Footbridge Modal Analysis

A modal analysis of the empty footbridge was undertaken to determine the modal mass, stiffness, damping, and mode shapes for the first several eigenfrequencies of the footbridge. The modal analysis was performed through a series of free-decay tests. For each test, a set of DC accelerometers were placed as reference accelerometers in the center of the bridge on either side and another set was moved along the bridge, from the center towards the supports. The accelerometers used were Bruel & Kjær 4575 with a 2 g range.

Fig. 4.3 Actuator at footbridge mid-span with hinged load cell

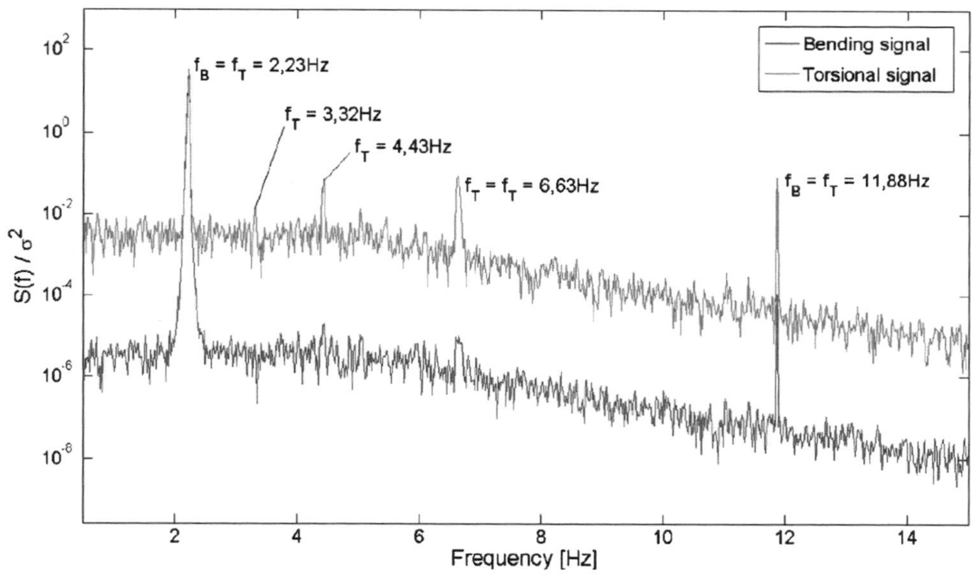

Fig. 4.4 Normalized PSD of bending and torsional signals measured during the footbridge modal analysis

Figure 4.4 shows the normalized power spectral density (PSD) of a set of bending and torsional accelerations measured during the modal excitation. From this, the footbridge frequencies can be identified, for both bending and torsional modes. A finite element analysis of the footbridge identified the same modes, with small deviations in frequency. The first bending eigenmode was found to be near perfectly sinusoidal in shape.

The modal frequency and damping were both found to be amplitude dependent. Figure 4.5 shows the dependency of frequency on vibration amplitude, whilst Fig. 4.6 shows the amplitude dependency of the damping of the empty bridge. The plotted points are a result of five independent vibration tests.

Frequency response curves were also established for the empty footbridge, once the actuator of Fig. 4.3 was installed, but no discernable change in the structures modal properties could be measured.

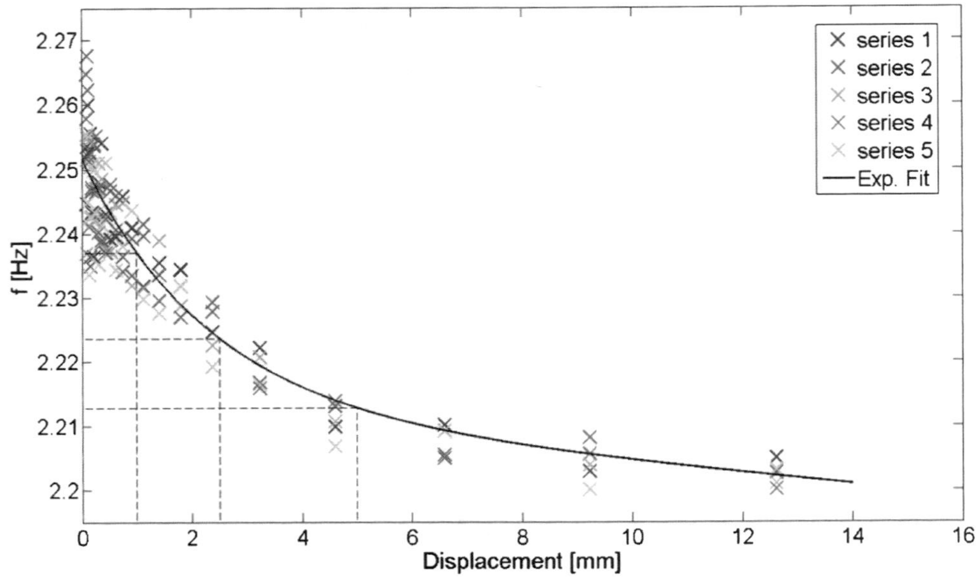

Fig. 4.5 Amplitude dependent first natural frequency for the empty footbridge

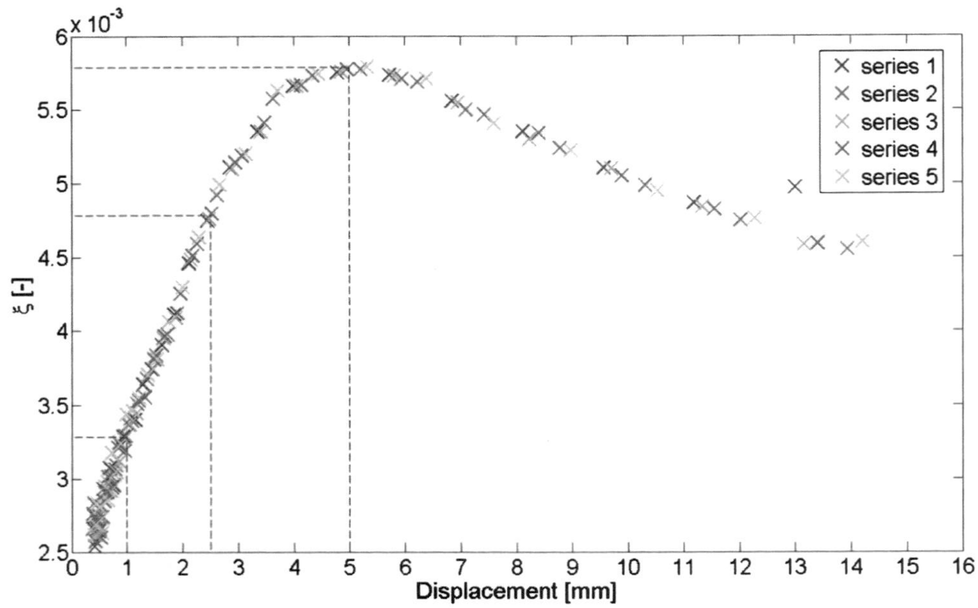

Fig. 4.6 Amplitude dependent damping for the empty footbridge

4.3.2 Pedestrian Tests

A series of Monte Carlo simulations were run to evaluate the minimum number of pedestrians and time that would be needed to ensure loading stationarity. The minimum number of pedestrians was found to be four, whilst the minimum test duration was found to be 180 s. During the tests, mean pedestrian flow rates were varied, but it was subsequently found that the change in footbridge mass and damping from each pedestrian was flow-independent. To vary the flow rates, tests were repeated with 4, 7 and 10 pedestrians, leading to mean flow rates of approximately $\lambda = 0.35$, 0.62 and 0.88 ped/s. Pedestrians were allowed to walk in both directions along the bridge.

A frequency sweep was undertaken for each distinct mean flow rate and mid-span amplitude, to determine the natural frequency of the footbridge with the moving pedestrians on it. Tables 4.1, 4.2 and 4.3 show the excitation frequencies for each flow rate, respectively.

Table 4.1 Excitation frequencies versus amplitude for $\lambda = 0.35$

Amplitude [mm]	Frequency [Hz]							
1	2.086	2.136	2.155	2.174	2.192	2.211	2.230	2.280
5	2.073	2.123	2.142	2.162	2.182	2.201	2.220	2.270
10	2.029	2.079	2.097	2.115	2.134	2.152	2.170	2.220

Table 4.2 Excitation frequencies versus amplitude for $\lambda = 0.62$

Amplitude [mm]	Frequency [Hz]							
1	2.022	2.072	2.104	2.135	2.167	2.198	2.230	2.280
5	2.009	2.059	2.091	2.123	2.156	2.188	2.220	2.270
10	1.965	2.015	2.046	2.077	2.108	2.139	2.170	2.220

Table 4.3 Excitation frequencies versus amplitude for $\lambda = 0.88$

Amplitude [mm]	Frequency [Hz]							
1	1.980	2.030	2.074	2.118	2.162	2.206	2.225	2.300
5	1.970	2.020	2.064	2.108	2.152	2.196	2.240	2.290
10	1.950	2.000	2.044	2.088	2.132	2.176	2.220	2.270

The distribution between male and female participants was 50/50 for the lowest and highest flow rates and 86/17 for $\lambda = 0.62$. Participants for the first two flow rates wore mainly comfortable athletic shoes (trainers). For the highest flow rate, flip-flops or heals were worn mainly, instead of trainers. The mean pacing frequency of $f_p = 1.38$ Hz for all the tests was much lower than might be expected for the Danish population, where f_p is closer to 1.83 Hz. This is mostly likely due to the test environment and the physical and psychological constraints imposed by the narrow deck without handrails.

4.4 Results

Intuitively, it might be expected that the average added mass per pedestrian is equivalent to the mass of each individual pedestrian. This is in fact what was found from the tests, with the average added mass per pedestrian being 102 % their total individual mass. This value was found to be constant, regardless of flow rate and vibration amplitude or frequency tested.

In contrast to what has been previously reported for vibrations in the lateral direction [6], the pedestrians here were found to always add damping vertically to the bridge when walking across it. The measured added damping had some level of scatter in it, due to the stochastic nature of the pedestrian loading. Therefore, a statistical analysis of the damping per cycle was performed to evaluate the probability of occurrence of varying magnitudes of added pedestrian-attributable damping. For all vibration amplitudes and flow rates, it was found that the Weibull distribution was a best fit to the data. The Weibull probability density function for the pedestrian damping, $P(c_p)$, can be written as:

$$P(c_p;\, \mu, \kappa) = \frac{\kappa}{\mu}\left(\frac{c_p}{\mu}\right)^{\kappa-1} e^{-(c_p/\mu)^{\kappa}} \tag{4.11}$$

where μ and κ are the scale and shape parameters of the distribution. The mean, standard deviation and fractile values of the distributions describing the pedestrian damping coefficients for each flow rate and mean floor displacement (above 1 mm) are presented in Tables 4.4, 4.5, and 4.6. The scale and shape parameters of the distributions are also provided.

An examination of the distributions revealed similarities in the pedestrian damping coefficients from one flow rate to another. Therefore, it was found that an exponential fit could be made to the data to reveal amplitude-dependent and flow-independent pedestrian damping coefficients, c_p, for varying fractile levels. The exponential fit is of the form:

$$c_p = a \cdot e^{bx} + c \cdot e^{dx} \tag{4.12}$$

where x represents the mean floor displacement in millimeters and a, b, c, d are coefficients that are determined based on a least square fit. The coefficients are provided in Table 4.7. Substitution of the coefficients of Table 4.7 into (4.12) leads to Fig. 4.7, in which the amplitude-dependent pedestrian damping coefficient is plotted for each fractile.

Table 4.4 Weibull distribution parameters describing the added pedestrian damping coefficients for $\lambda = 0.35$

Mean amplitude [mm]	10 % [kg/s]	50 % [kg/s]	90 % [kg/s]	95 % [kg/s]	c_p mean [kg/s]	σ [kg/s]	μ [−]	κ [−]
1.6	201.5	1381.8	3643.5	4472	1706.3	1339.3	2056.3	1.343
3.2	40.7	770.2	2145.6	2645.7	966.3	817.1	1274.7	1.364

Table 4.5 Weibull distribution parameters describing the added pedestrian damping coefficients for $\lambda = 0.62$

Mean amplitude [mm]	10 % [kg/s]	50 % [kg/s]	90 % [kg/s]	95 % [kg/s]	c_p mean [kg/s]	σ [kg/s]	μ [−]	κ [−]
1.6	197.2	1113.3	2903.5	3565.2	1366.8	1112	1590.8	1.318
3.2	65.9	685.9	1930	2395.4	875.29	716.6	1072.4	1.287

Table 4.6 Weibull distribution parameters describing the added pedestrian damping coefficients for $\lambda = 0.88$

Mean amplitude [mm]	10 % [kg/s]	50 % [kg/s]	90 % [kg/s]	95 % [kg/s]	c_p mean [kg/s]	σ [kg/s]	μ [−]	κ [−]
1.7	207.3	1083	2862.6	3532.3	1347.7	1055.6	1511.8	1.272
3.2	72.2	593.3	1643.5	2037.3	754.5	605.1	900.6	1.281

Table 4.7 Coefficients for the exponential fit of (4.12)

	fractile			
	10 %	50 %	90 %	95 %
a	2224.9	17,457	72,456	98,698
b	−3.54	−4.49	−5.42	−5.05
c	622.08	2026.3	5101.5	6231.91
d	−0.76	−0.35	−0.32	−0.32

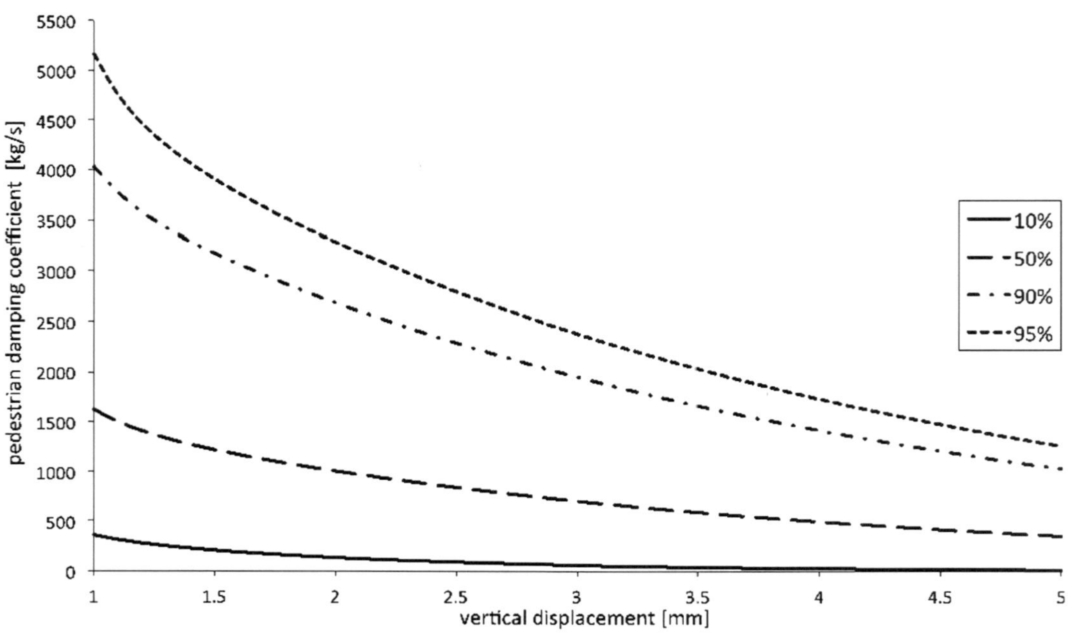

Fig. 4.7 Amplitude dependent pedestrian damping coefficients for varying probability fractiles

As only three distinct vibration amplitudes were tested, it should be reasonable to suggest that the exponential fit of (4.12) could be substituted with a linear fit, which might also be more conservative for design purposes for lower vibration amplitudes. Finally, Fig. 4.7 reveals a reduction in pedestrian damping with amplitude of vibration. This indicates a change in the way pedestrians interact with a footbridge for larger vibrations.

For design purposes, a pedestrian may be treated as a moving point viscous damper, in which case a conservative fixed value of $c_p = 500$kg/s is suggested for moderate vertical vibrations of up to 5 mm amplitude.

4.5 Conclusions

The added mass and damping introduced by pedestrians on a vertically vibrating footbridge has been determined through a series of forced vibration tests on a 16 m long laboratory footbridge. For all amplitudes of vibration and pedestrian flow rates, the added mass of a single pedestrian was found to be 102 % of the pedestrian's mass. The damping introduced by each pedestrian was found to be positive for all amplitudes, but decreasing with increasing amplitudes of vibration. A conservative fixed value of 500 kg/s is suggested as an individual pedestrian damping coefficient for design purposes.

References

1. SETRA, Footbridges, Assessment of vibrational behavior of footbridges under pedestrian loading (2006) Technical guide, Technical Department for Transport, Road Bridges, Engineering and Road Safety, Paris, Oct 2006
2. FIB (2005) Guidelines for the design of footbridges, bulletin 32, Fédération international du béton (fib), Nov 2005
3. Georgakis CT, Ingólfsson ET (2008) Vertical footbridge vibrations: the response spectrum methodology. In: Proceedings of footbridge 2008, Porto, 2–4 July 2008
4. Živanović S, Díaz IM, Pavić A (2009) Influence of walking and standing crowds on structural dynamic properties. In: Proceedings of IMAC-XXVII, Orlando, 9–12 Feb 2009
5. Clough RW, Penzien J (1975) Dynamics of structures, 1st edn. McGraw-Hill, New York
6. Ingólfsson ET, Georgakis CT, Ricciardelli F, Jönsson J (2011) Experimental identification of pedestrian-induced lateral forces on footbridges. J Sound Vib 330:1265–84

Chapter 5
Predicting Footbridge Response Using Stochastic Load Models

Lars Pedersen and Christian Frier

Abstract Walking parameters such as step frequency, pedestrian mass, dynamic load factor, etc. are basically stochastic, although it is quite common to adapt deterministic models for these parameters. The present paper considers a stochastic approach to modeling the action of pedestrians, but when doing so decisions need to be made in terms of statistical distributions of walking parameters and in terms of the parameters describing the statistical distributions. The paper explores how sensitive computations of bridge response are to some of the decisions to be made in this respect. This is useful as it pinpoints which decisions to be concerned about when the goal is to predict footbridge response. The studies involve estimating footbridge responses using Monte-Carlo simulations and focus is on estimating vertical structural response to single person loading.

Keywords Probabilistic modeling • Footbridge vibrations • Walking loads • Sensitivity study • Vibration comfort

Nomenclature

f_0	Bridge frequency
l_s	Stride length
q	Modal load
k	Response ratio
ζ	Bridge damping ratio
f_s	Step frequency
m	Weight of pedestrian
L	Bridge length
A	Acceleration property
μ	Mean value
F	Walking load
P	Prob. density function
M	Bridge modal mass
α	Dynamic load factor
σ	Standard deviation

L. Pedersen (✉) • C. Frier
Department of Civil Engineering, Aalborg University, Sohngaardsholmsvej 57, Aalborg DK-9000, Denmark
e-mail: lp@civil.aau.dk

A. Cunha (ed.), *Topics in Dynamics of Bridges, Volume 3: Proceedings of the 31st IMAC, A Conference on Structural Dynamics, 2013*, Conference Proceedings of the Society for Experimental Mechanics Series 38, DOI 10.1007/978-1-4614-6519-5_5, © The Society for Experimental Mechanics, Inc. 2013

5.1 Introduction

The walking-induced vibrations in the London Millennium Bridge [1] serve as an example of a scenario where walking loads have caused serviceability problems. However, it is by far the only footbridge, which has shown to be problematic in the serviceability-limit-state, but probably it is the most well-known bridge, that has had problems in this area.

When addressing the serviceability-limit-state, many codes of practise employ a deterministic approach in which the walking parameters influencing the walking load is handled as deterministic properties (such as in [2, 3]). The weight of a pedestrian is set to 75 kg, the dynamic load factor for a pedestrian is set to 0.4, etc. In [4] a new way of thinking (modelling) was introduced in which at least some of the walking parameters were modelled as random variables rather than deterministic properties. This is more reasonable as walking parameters are stochastic by nature as documented in [4–6]. At the same time the load modelling procedure is more complex, as there are many decisions to be made prior to launching calculations predicting the footbridge response. These decisions encompass settling on mean values, standard deviations, distribution type etc. for the various walking parameters.

In the paper focus is on setting up such assumptions for two main characteristics describing the load action, namely the pedestrian weight and the dynamic load factor. Furthermore, and quite important, it is examined how sensitive the footbridge response actually is to the decisions made along the way.

Overall the idea is to get an understanding of how the various decisions impact the predictions of footbridge response, or in other words how sensitive footbridge response predictions are to the various decisions to be made. One study of this paper is devoted to decisions about the dynamic load factor. Another is devoted to decisions about the pedestrian weight.

To facilitate the investigations, footbridge models are needed, and to this end pin-supported footbridges (idealised as a single-degree-of-freedom systems, SDOF-systems) are employed. Actually five different pin-supported bridges are considered, such that the results of sensitivity studies do not only reflect results obtained for a single randomly selected bridge.

The bridge response characteristic in focus is defined in the paper.

Dynamic characteristics of the bridges considered in this paper are outlined in Sect. 5.2, and the way in which walking loads are modelled is outlined in Sect. 5.2. Section 5.3 and 5.4 presents the sensitivity studies and Sect. 5.5 summarizes the results.

5.2 Bridge Dynamic Characteristics

Five different pin-supported footbridges are considered for the studies of this paper. The dynamic characteristics of the bridges are shown in Table 5.1. The parameter f_0 represents the undamped bridge frequency (first vertical bending mode), M the associated modal mass, and ζ represents the damping ratio of the bridge.

Basically, bridge length, L, is not a dynamic characteristic as such. However, as it is a parameter having impact on bridge response to the action of walking, it is a parameter, which need to be defined for a study like this.

Bridge C would have the highest risk of resonating as a result of walking loads, as for this paper the mean value of step frequency is assumed to be 1.87 Hz, but the other bridges might also be brought into resonance as a result of walking loads although it is less likely. Bridge A and E have frequencies in the lower and upper frequency range of natural walking, respectively, which would suggest that resonance action will not occur that often, but that it is a possibility.

All together, the range of bridges outlined above is believed to be suitable for the parametric studies of this paper. A variability of pin-supported bridges is considered and it is also believed that there is a fair relationship between bridge frequency, modal mass and bridge length.

Table 5.1 Bridge dynamic characteristics

Bridge	f_0 (Hz)	M (kg)	ζ (%)	L (m)
A	1.60	$61.5 \cdot 10^3$	0.5	54
B	1.75	$51.5 \cdot 10^3$	0.5	49
C	1.90	$44.0 \cdot 10^3$	0.5	45
D	2.05	$37.5 \cdot 10^3$	0.5	42
E	2.20	$32.5 \cdot 10^3$	0.5	39

5.3 Modelling of Walking Loads

When modelling walking loads, Fourier series are often employed, but for the present studies, the basic excitation frequency (step frequency f_s) is so close to matching the bridge frequency, that it is not considered necessary to consider super-harmonic load components. Basically, because the super-harmonic load components are not likely to be able to cause significant resonant effects.

Therefore the vertical dynamic load induced by a pedestrian is modelled according to (5.1):

$$f(t) = mg\alpha \cos(2\pi f_s t) \tag{5.1}$$

in which t is time, and m is pedestrian weigth (in kg). The parameter g represents acceleration of gravity (for simplicity $g = 10$ N/kg is assumed), but it is also apparent that the parameter α (non-dimensional), and called the dynamic load factor, has an impact on the amplitude of walking loads. For reference it is mentioned that the modelling approach is in agreement with [7–9].

When employing the model for walking load introduced above, it is assumed that the pedestrian walks with a constant step frequency, f_s, while crossing the bridge. Nevertheless, the paper will model it so that the step frequency may vary from one bridge crossing to the next hereby seeking to embrace the random nature of the step frequency known to exist.

On these assumptions, the modal load acting on the bridge, $q(t)$, may be computed using (5.2):

$$q(t) = mg\alpha \sin(2\pi f_s t)\ \Phi(t) \tag{5.2}$$

where $\Phi(t)$ is the value of the mode shape function for the first vertical bending mode of the footbridge at the current position of the pedestrian. This value may be computed using (5.3):

$$\Phi(t) = \sin\left(\frac{\pi f_s l_s}{L} t\right) \tag{5.3}$$

as the walking velocity v may be derived from the following relationship:

$$v = f_s\, l_s \tag{5.4}$$

in which l_s, represents the stride length of the pedestrian. For later computations, it is assumed that any pedestrian traverses the bridge using a constant stride length, l_s, (step length), and a constant step frequency, f_s, and thus with a constant walking speed v. Nevertheless, the paper will model it so that the step frequency and the stride length may vary from one bridge crossing to the next hereby seeking to model the random nature of both properties (and thereby a random nature of walking velocity).

Having outlined the basic premises for the studies, more details are now given for the specific sensitivity studies of this paper.

5.4 Study Related to Modelling the Dynamic Load Factor

It is so that there are various ways to model the dynamic load factor. For the sensitivity studies of this paper, three different potential models are considered and they are outlined in Sect. 5.4.1. Section 5.4.2 supplements with other study assumptions, and the last section discusses the results.

5.4.1 Dynamic Load Factor

Model 1 In this model, the dynamic load factor is modelled to depend on step frequency, f_s, according to the relationship given in (5.5) [6]:

$$\alpha = c_1 f_s^3 + c_2 f_s^2 + c_3 f_s + c_4 \tag{5.5}$$

Table 5.2 Mean values (μ) and standard deviations (σ)

Model	Unit	μ	σ
1	[−]	(5.5)	(5.7)
2	[−]	(5.5)	0
3	[−]	(5.8)	0

Table 5.3 Mean values (μ) and standard deviations (σ)

Variable	Unit	μ	σ
f_s	Hz	1.87	0.186
l_s	m	0.71	0.071
m	kg	75	13.5

where

$$c_1 = -0.2649 \quad c_2 = 1.306 \quad c_3 = -1.7597 \quad c_4 = 0.7613. \tag{5.6}$$

In (5.5), α represents the mean value of the dynamic load factor for a given step frequency, f_s [inserted in Hz in (5.5)]. According to [6], the standard deviation of the dynamic load factor (denoted σ_α) may be computed using (5.7):

$$\sigma_\alpha = 0.16\,\mu_\alpha \tag{5.7}$$

In this model for the dynamic load factor a random nature of the load factor is assumed, and the load factor is modelled as a Gaussian distribution with a mean value defined by (5.5) and a standard deviation defined by (5.7). It is believed that there is good reasoning in choosing a Gaussian distribution as data sets in [6] generally supported this assumption.

Model 2 In model 2, the standard deviation is set to zero, hereby handling the dynamic load factor as a deterministic property. Its value attains the property found by using (5.5). Introducing this model facilitates comparisons to be made with the model 1 in which another assumption/decision was made for standard deviation.

Model 3 Another proposal as to how to model the dynamic load factor may be found in [7]. In [7] only the raw data measuring results for the dynamic load factor are provided, so no formula is given, but the authors of the present paper has made a curve-fit to the data, and found the following relationship:

$$\alpha = b_1 f_s^5 + b_2 f_s^4 + b_2 f_s^2 + b_3 f_s + b_4 \tag{5.8}$$

where

$$b_1 = 4.1407 \quad b_2 = -43.3815 \quad b_3 = 181.0209 \quad b_4 = -376.6609 \quad b_5 = 391.8452 \quad b_6 = -163.16. \tag{5.9}$$

The measuring results also displayed some scatter, but the scatter will not be considered for the studies of this paper, hence the standard deviation will be set to zero. Hereby, model 3 represents a deterministic model for the dynamic load factor, but a model different from that introduced as model 2.

In summary, the following models for the dynamic load factor are considered (Table 5.2).

The model numbers will be used later to trace the assumptions made.

5.4.2 *Other Study Assumptions*

As concerns the other properties involved in modelling the walking load, their mean values and standard deviations are given in Table 5.3.

On these assumptions, step frequency, stride length and pedestrian weight are modelled as random variables and Gaussian distributions are assumed for each variable. The mean value of the pedestrian weight is set to 75 kg simply because, it is a value occasionally assumed for pedestrians, and the standard deviation is set to a value that reasonably well matched that found from a sample of weights of students at Aalborg University. However, the population was not that high, so it can be discussed whether at is in fact a reasonable value.

The mean values and standard deviations employed for step frequency and stride length comply with values suggested in [4].

Table 5.4 Ratio $k(y)$

Bridge	A	B	C	D	E
$k(1)$	1	1	1	1	1
$k(2)$	~1	~1	~1	~1	~1
$k(3)$	~1	~1	1.1	1.15	1.15

5.4.3 *Computations*

A Newmark time integration scheme was employed to determine a bridge acceleration time history at bridge midspan each time a pedestrian crossed the bridge. MonteCarlo-simulations ensured that the random nature of the load was modelled according to the assumptions, and it was ensured that the number of simulation runs (bridge crossings) were higher than what was actually required to give sound estimates of the bridge acceleration property in focus.

The acceleration property in focus was the peak acceleration in the time series (such a value was extracted for each bridge crossing) and focus was on the 95 % quantile of all identified peak accelerations. This value was extracted from the probability distribution function that is the first result of the simulations. For simplicity, the 95 % quantile will hereafter be denoted a, and will be referred to as the acceleration property. This property (the 95 % quantile) has previously been the focus of interest in studies dealing with bridge response in a stochastic manner [4, 10], and is therefore of interest to monitor. By definition the value, a, is expected to be exceeded in 1 out 20 bridge crossings.

5.4.4 *Results*

Values of the acceleration property a were computed for all five bridges on the three different assumptions for modelling the dynamic load factor (for model y, where y represent the model number being either 1, 2 or 3). For the presentation of results for the five bridges, it is useful to introduce the ratio k defined in (5.10).

$$k(y) = a(y)/a(1) \tag{5.10}$$

The ratio relates acceleration properties calculated assuming model y ($a(y)$) to those obtained when employing model 1 ($a(1)$), in which the dynamic load factor was modelled as a random variable.

Results in term of the ratio k, is given in Table 5.4 for the five bridges (A-E). For ease of overview only approximate values for the ratio k is given.

In Table 5.4, the notation ~1 indicates that there is a marginal difference from the value of 1 (at least less than a 3 % difference). Of cause and by definition, the ratio $k(1)$ attains a value of 1, but it is interesting to observe that the ratios $k(2)$, and for all bridges, attain values very close to 1. The only difference between model 1 and model 2 is that in model 2, the standard deviation of the dynamic load factor is set to zero. Hence, the results suggest that the acceleration property a is fairly insensitive to whether the standard deviation is modelled. In other words the acceleration property a is fairly insensitive to whether the dynamic load factor is modelled as a random variable (with non-zero standard deviation) or as a deterministic property.

Another thing that can be seen in Table 5.4 is that for the bridges C, D, and E, model 3 results in a higher value of a than model 2. Both model 2 and 3 are characterised by having the standard deviation of the dynamic load factor set to zero. Hence the only difference between the two models is the mean value curve for the dynamic load factor [(5.5) and (5.8), respectively]. If these were plotted, it would appear that for frequencies above 1.75 Hz, the dynamic load factor is highest in model 3. This directly explains the observation in that the bridges C, D and E have frequencies higher than 1.75 Hz.

Overall the results suggest that the acceleration property a is fairly insensitive to the standard deviation of the dynamic load factor but it is sensitive to the mean value (curve) of the dynamic load factor. Hence, it is important to employ a representative mean value (curve) for the dynamic load factor for bridge response calculations.

5.5 Study Related to Modelling Pedestrian Weight

Some uncertainty exists when it comes to modelling the weight of the pedestrian (as already indicated in the previous section). In that section a Gaussian distribution was assumed, and this approach is also employed here as a first estimate. However, in the previous section only a single value was assumed for the mean value of pedestrian weight and a single accompanying value (non-zero value) was assumed for the standard deviation of pedestrian weight.

Table 5.5 Mean values (μ) and standard deviations (σ), $n = 0$, 0.1,0.2, 0.3

Variable	Unit	μ	σ
m	kg	50	$n\mu$
m	kg	75	$n\mu$
m	kg	85	$n\mu$

This hardly represents the possible combinations on a footbridge, and therefore a more detailed study is made to bring light onto how sensitive the acceleration property a is to different assumptions for the pedestrian weight. This involves considering a total of 12 combinations of mean values and standard deviations for pedestrian weight ranging from models in which the pedestrian weight is treated as a random variable, to models where it is treated as a deterministic property.

The considerations (variation possibilities) may quickly expand, and therefore focus is only on bridge C, being the bridge for which resonant excitation is most likely to occur.

5.5.1 *Weight of Pedestrian*

As for the pedestrian weight, three different mean values were considered; a low, a medium, and a high value (50, 75 and 85 kg) as shown in Table 5.5.

Table 5.5 also defines the standard deviations that are assumed associated with the mean values, and for each mean value four different standard deviations are considered. With $n = 0$, the standard deviation is set to zero hereby handling pedestrian weight as a deterministic parameter. However, three other options for non-zero standard deviation is also considered ($n = 0.1$, 0.2, and 0.3) all modelling pedestrian weight as a random variable are considered. As n increases, so does the standard deviation (from low over medium to high). The standard deviation introduced in Sect. 5.3 is fairly close to assuming $n = 0.2$ hereby giving some idea of the standard deviation ranges considered for the present studies.

5.5.2 *Other Study Assumptions*

The step frequency and stride length was modelled as in Sect. 5.3. In Sect. 5.3 three different models for the dynamic load factor were introduced, but for the present studies model 1 was employed.

5.5.3 *Computations*

Basically, the computations followed the same scheme as that already outlined in Sect. 5.4.3. On various study assumptions the acceleration property in focus in this paper was identified, hereby allowing assessments to be made as to how sensitive the acceleration property is to variations in pedestrian weight modelling.

5.5.4 *Results*

The results are presented by employing the ratio k defined in (5.11):

$$k(m,n) = a(m,n)/a(75,0) \tag{5.11}$$

Basically, the ratio normalises the acceleration property, a, calculated for any studied combination of (m,n) by the acceleration property, a, calculated for $(m, n) = (75 \text{ kg}, 0)$. The normalisation constant represents the acceleration property calculated assuming the pedestrian weigh to be 75 kg and a deterministic model for the pedestrian weight (as $n = 0$). The normalisation is believed to enhance the interpretation of results. As n grows above zero, the pedestrian weight is handled as a random variable, and as n increases, the standard deviation of pedestrian weight increases.

The results in terms the ratios k, are given in Table 5.6. For ease of overview, only approximate ratios are provided.

Table 5.6 Ratio $k(m, n)$

$m \backslash n$	0	0.1	0.2	0.3
50	0.67	0.67	0.67	0.68
75	1	1	1	1.02
85	1.13	1.13	1.13	1.16

One item to notice is that for $m = 5$ kg, the ratio attains almost similar values regardless of the value of n. This indicates that it is not that important to model m as a random variable. Another item to notice is that values are also almost similar for any ($m = 50$ kg, n)-combination and any ($m = 85$ kg, n)-combination. This supports that the conclusion just made is not one only valid for $m = 75$ kg. Apparently, it is valid for low, medium, and high pedestrian weight scenarios.

However, it can also be inferred from Table 5.6, that the acceleration property increases, when m increases. This would also be expected, as an increased pedestrian weight increases bridge acceleration levels.

The overall conclusion seems to be that for a reasonable estimate of the acceleration property a, it is more important to decide on a reasonable value of m than spending time on deciding on a reasonable value of n.

Although it is not shown here, it can be shown that this conclusion is also valid if one should decide to entirely shift the distribution type from the assumption made here (a Gaussian distribution) to a log-normal distribution for pedestrian weight.

5.6 Conclusion

The results of the sensitive studies imply that the bridge acceleration property in focus in this paper (the 95 % quantile of midspan peak bridge accelerations) is relatively insensitive to the values of standard deviations assumed for the pedestrian weight and for the dynamic load factor.

This might suggest that it would not be necessary to model these properties as random variables. The bridge acceleration property, on the other hand, is to a high degree dependant on the mean values assumed for these properties.

There would be limits as to the variability of the mean value of the dynamic load factor, but the variability of the mean value of pedestrian weight is probably much higher making this a parameter of particular interest to study further.

Without the sensitivity studies, these conclusions would be difficult to foresee.

However, it must be recalled that the sensitivity studies only considered the response of pin-supported bridges, and that multi-person pedestrian traffic was not addressed in the studies.

References

1. Dallard P, Fitzpatrick AJ, Flint A, Le Bourva S, Low A, Ridsdill-Smith RM, Wilford M (2001) The London Millennium Bridge. Struct Eng 79:17–33
2. Ontario Highway Bridge Design Code, Highway Engineering Division (1983) Ministry of Transportation and Communication, Ontario
3. British Standard Institution (1978) Steel, concrete and composite bridges. Specification for loads. British Standard Institution, London, BS 5400: Part 2
4. Živanovic S (2006) Probability-based estimation of vibration for pedestrian structures due to walking. Ph. D. thesis, Department of Civil and Structural Engineering, University of Sheffield, Sheffield
5. Matsumoto Y, Nishioka T, Shiojiri H, Matsuzaki K (1978) Dynamic design of footbridges. In: IABSE proceedings, No. P-17/78, Zürich, pp 1–15
6. Kerr SC, Bishop NWM (2001) Human induced loading on flexible staircases. Eng Struct 23:37–45
7. Ellis BR (2000) On the response of long-span floors to walking loads generated by individuals and crowds. Struct Eng 78:1–25
8. Bachmann H, Ammann W (1987) Vibrations in structures – induced by man and machines. IABSE Structural Engineering Documents 3e, Zürich
9. Rainer JH, Pernica G, Allen DE (1998) Dynamic loading and response of footbridges. Can J Civ Eng 15:66–78
10. Pedersen L, Frier C (2009) Sensitivity of footbridge vibrations to stochastic walking parameters. J Sound Vib. doi:10.1016/j.jsv.2009.12.022

Chapter 6
Temperature and Traffic Load Effects on Modal Frequency for a Permanently Monitored Bridge

Yavuz Kaya, Martin Turek, and Carlos Ventura

Abstract Many of the damage detection techniques available compare measured parameters of a structure in a known, reference state to an unknown, possibly damaged state. In the case where the measured parameters are the modal characteristics of the structure, it has been shown that these can be sensitive to changes other than damage, such as temperature. Therefore it is useful to install permanent monitoring systems that can track normal changes in measured modal characteristics. The Ironworkers Memorial Second Narrows Crossing, in Vancouver, Canada, had a permanent monitoring system installed in 2011. The purpose of the system is to monitor the response of the bridge during seismic and other potentially damaging events, and to perform long term performance monitoring. During an assessment of the modal parameters it was observed that there is a regular fluctuation in the measured frequencies, so a detailed study on a 17-day segment of continuous data was performed. Analysis was performed on both temperature data and acceleration amplitude (to represent traffic on the bridge). The modal fluctuations were very regular, with maximum at night and minimums in the day, which correlated well with the maximum traffic in the day and minimums at night. The temperature fluctuations coincided with daily changes but had a more variable pattern, which was not directly correlated to changes in frequency. It is also observed that the traffic load on the bridge amplifies the response of the first vertical mode seven and ten times during daytime weekdays and daytime weekends, respectively whereas the first torsional mode of the structure amplifies 5.5 and 4 times in daytime weekdays and weekends, respectively.

Keywords Modal identification • Traffic load • Temperature effect • RMS value • Average amplification

6.1 Introduction

The Ministry of Transportation and Infrastructure (MoT) and the University of British Columbia (UBC) have initiated a seismic and Structural Health Monitoring (SHM) program of bridges in the province of British Columbia. The purpose of the program is to monitor the response of the bridges during severe events, such as seismic, wind, impact; and to continuously monitor the day-to-day performance. As part of the system analysis, an objective will be to detect damage of the structures using field data and state-of-the-art damage detection algorithms. Many of the damage detection algorithms estimates and locates damage on a structure by comparing a reference state, usually intact structure, with the most current estimated state, possibly damaged one. The estimation of current structural state can be done based on the modal characteristics of the structure (e.g. modal frequency, damping ratio, and mode shape). However, it is well known that environmental conditions such as temperature and traffic loads can have significant effect in the modal characteristics of the structure.

Temperature affects the Young's modulus of steel and concrete [1, 2], which in turn changes the boundary conditions, and then the modal properties of the structure [3]. Traffic load acting on the bridge changes the mass of the structure and may affect the structure's modal properties [4]. Changes in natural frequency in the order of 10 % or more from environmental sources have been observed [5, 6]. The analyses of six different sets of earthquake records, as well as

Y. Kaya (✉) • C. Ventura
The University of British Columbia, Vancouver, Canada
e-mail: kayaya@mail.ubc.ca; ventura@civil.ubc.ca

M. Turek
British Columbia Ministry of Transportation and Infrastructure, Victoria, Canada

A. Cunha (ed.), *Topics in Dynamics of Bridges, Volume 3: Proceedings of the 31st IMAC, A Conference on Structural Dynamics, 2013*, Conference Proceedings of the Society for Experimental Mechanics Series 38,
DOI 10.1007/978-1-4614-6519-5_6, © The Society for Experimental Mechanics, Inc. 2013

ambient vibration records, from a 40-story steel building in Los Angeles have shown that, although there was no damage, the natural frequencies of the building changed as much as 30 % due to nonlinearities in the building's response and the soil-structure interaction effects [7]. By analyzing 2-year-long continuous records from the Caltech's Millikan Library, the natural frequencies of a building can change significantly due to environmental factors, such as rainfall, wind speed, and temperature [8]. In an analytical investigation of a 10-story building, it has been shown that in order to see a 10 % reduction in the fundamental frequency, more than 40 % reduction in one of the story stiffnesses is required [7]. Such a large reduction in stiffness would normally cause clearly visible damage.

Thus this indicates a challenge for SHM-based damage detection systems because the change of modal properties of the structure due to environmental conditions can be larger than the change due to the damage. Therefore the effect of environmental conditions (traffic load in case of bridge) on modal properties of structure must be accounted for as they can completely mask the change of modal properties caused by real damage. Researchers have studied this effect and found different relationships for different structures. For instance, the influence of temperature on modal frequencies and their correlation have been investigated for a cable stayed bridge in Hong Kong using long-term monitoring data where it has been found that the environmental temperature changes account for variation in modal frequencies [9]. In a different study using linear adaptive models, changes in the frequencies are found linearly correlated with temperature readings from different parts of the Alamosa Canyon Bridge in New Mexico in USA [10]. A study about the correlation between modal properties and the temperature showed that modal frequencies have a clear negative correlation with temperature and humidity while damping ratios have a positive correlation, but no clear correlation of mode shapes with either temperature or humidity [11].

Such effects should be accounted for before applying any damage detection algorithm, which is based on the change in modal characteristics of structure. The effect of such environmental conditions must be removed from the estimated modal characteristics of the structure in order to have more accurate results in damage detection algorithms. One of the primary tools to deal with this issue is in long term monitoring of a structure, where the modal characteristics can be catalogued over time, and any changes due to normal conditions can be observed.

Recently, a remotely-accessible, permanent seismic and SHM system was installed on the Ironworkers Memorial Second Narrows Crossing (IMSNC) by the Ministry of Transportation and Infrastructure of British Columbia. The data are continuously monitored, processed and stored in a local database; the processed data and reports are then transferred to a central database and posted to the internet, which is part of the Provincial BCSIMS system [12]. Previously, a full ambient vibration test and model updating was done on this bridge [13]. The updating was done using a reference set of modes; however, with data from the permanent system, changes were observed in those reference values. Further examination showed that the changes appeared to be periodic.

This paper describes a study examining 17 days of data, and it was observed that the fundamental mode of the bridge (vertical in the main span) would change in a regular way increasing and decreasing daily. The two parameters which are suspected to affect this change are traffic load and temperature. Therefore the change in frequency is compared to daily local temperature change, and to the change in amplitude of vibration in the vertical direction at main span.

6.2 Description of the IMSNC

The IMSNC is part of an essential transportation corridor along the Trans-Canada highway (Route 1), connecting the City of Vancouver to the District of North Vancouver, the City of North Vancouver, and West Vancouver. It is a 1,290 m-long composite structure. It carries six lanes of traffic (three lanes for each direction) and a 0.6 m-diameter gas pipeline. The general configuration of the bridge and bent designation is illustrated in Fig. 6.1. This paper resents the results of a suspended span (between Sects. 15 and 16) only. For detailed information on the entire bridge see [13].

There are four 85.8 m-long steel approach truss spans starting on the North shore of the crossing and extending out into the inlet. Each span has two 13.0 m-depth custom steel warren trusses separated 14.6 m apart. Both trusses are connected by horizontal and vertical bracing. The 0.2 m-thick RC deck slab is supported on thirteen 0.8 m-depth steel I-girders in the longitudinal direction and on a 1.7 m-depth $\times$ 22.0 m-long steel I-floor beam located transversally every 10.5 m along the truss. The longitudinal girders are also connected transversally every 3.5 m by a 0.5 m-depth steel I-girders. The warren trusses are supported on isolation bearings protected with bumpers. The deck has transverse expansion joints at the ends of each span, so there is no structural continuity among them. The transition between two approaches and between the southernmost approach and the main cantilever section of the bridge is shown in Fig. 6.2.

The main bridge is made of up two anchor spans, cantilever spans and suspended spans. The anchors spans each have a massive tie down bar to a reaction beam below in the abutment (at the South end) and in Pier 14 at the North end. The anchor spans connect to cantilever spans, and pivot on large sliding bearings over Piers 15 and 16. Between the two cantilever spans is a 100 m suspended span; the span is pinned at both ends of the top chord, and pinned at one end and allowed to slide at one end of the bottom chord.

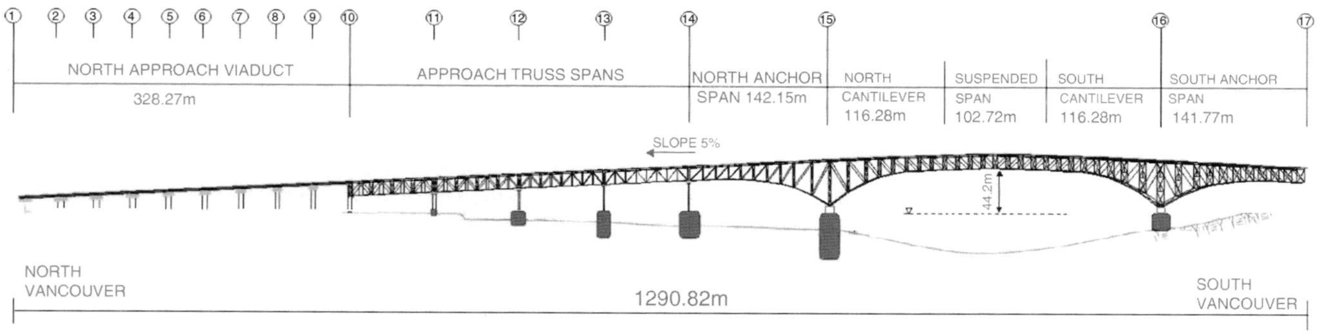

Fig. 6.1 General configuration and bent designation of the IMSNC

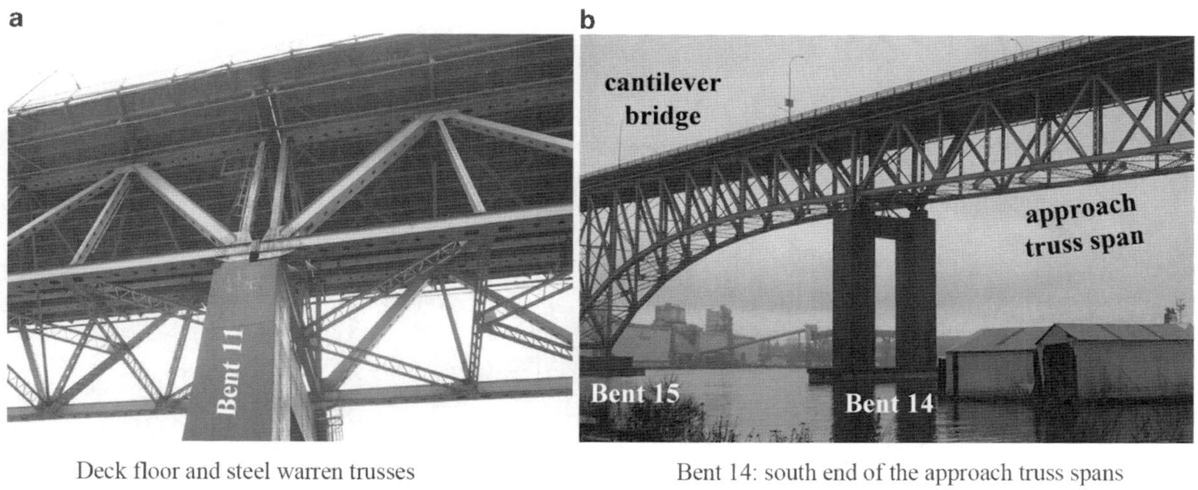

Fig. 6.2 Approach truss spans, (**a**) Deck floor and steel warren trusses, (**b**) Bent 14: south end of the approach truss spans

Since the original construction of the bridge, two significant seismic retrofit projects have been completed. The first was applied in 1994; this was a structural retrofit which involved the piers on the viaduct, strengthening both at the base and at the pier cap; the steel truss sections where several details were upgraded including adding isolation bearings and bumpers between individual truss spans, and at the anchor span tie-downs where the maximum longitudinal clearance at the top of the pier and abutment was increased. The second retrofit was applied in 2001 involving ground improvements on the north end of the bridge. This retrofit in general involved installing drains, ground densification and installing timber compaction piles.

6.3 Real-Time Seismic Monitoring System

A real-time monitoring system (RTMS) was installed on the IMSNC over several months, completed in May 2011. The monitoring system is a part of a province wide network, with its central hub located at the University of British Columbia in Vancouver. The monitoring system provides real-time information regarding structural performance and safety, primarily for seismic, but also applicable for a variety of load types. The monitoring system will be implemented in two phases: first to install the on-site hardware and second to implement a customized software and data processing system unique to the MoT/UBC network.

The general purpose of the system is to monitor the structural health of the bridge for seismic, impact and deterioration effects. This considers two loading levels: severe infrequent events, such as seismic and impact/collision; and frequent long-term effects, such as wind, traffic, etc. The system instrumentation consists of:

1. Tri-axial accelerometers on pier footings using 4 g force-balance EpiSensors
2. Tri-axial accelerometer at each abutment (2 in total) using 4 g force-balance EpiSensors

3. Strong motion measurements off the structure, including free-field and down-hole accelerometers using 4 g force-balance EpiSensors
4. Strain measurements on the deck floor beams and major truss elements using 2,000 $\mu\varepsilon$ dynamic gauges
5. Temperature measurements at several locations, both the North and South ends using thermistors
6. Wind speed measurements at midspan using a cup and vane sensor

The data will be collected at a central data recorder, in which a certain amount of on-site data processing will occur. Then processed and raw data will be sent to UBC for further processing and storage. The monitoring system has approximately 100 acceleration recording channels; however due to the configuration of isolation bearings and expansion joints, the bridge is essentially split into 10 smaller independent structures. This means that in reality there is an average of about 10 channels per structure.

6.3.1 SHM General Overview

The SHM system has been installed on the IMSNB as of May 2011. The monitoring system will be part of the province-wide BCSIMS (BC Smart Infrastructure Monitoring System) network, with its central server located at the UBC. The purpose of the system is to monitor the structural health of the bridge for seismic, impact and deterioration effects. This system considers two loading levels: severe infrequent events, such as seismic and impact/collision; and frequent long-term effects, such as wind, traffic, etc.

The BCSIMS network will consist of four elements: a webpage interface for both Authorized users and the general public (www.bcsims.ca); a main server to coordinate information from all of the monitored sites and communicate information to the website (SIMS1); a remote site computer to process data, make decisions on structural status, and to communicate to the central server (SIMS2); and lastly specialized data analysis servers to handle advanced processing (SIMS3). More information about BCSIMS can be found in [12].

6.3.2 SHM Details

For the viaduct section (from the North Abutment to the last concrete span between Piers 9 and 10):

- Tri-axial accelerometers at ground level at each of the North abutment, Piers 3, 5, 7 and 9.
- Two uni-axial accelerometers measuring transverse and longitudinal motions at the west side of pier caps on Piers 3, 5, 7 and 9

For each approach truss (10, 11, 12 and 13)

- Uni-axial accelemeters measuring longitudinal motion at the west side, top chord, at each end
- Uni-axial accelerometer measuring Vertical motion at top chord, midspan, east and west sides;
- Uni-axial accelerometer measuring Transverse motion at top chord, midspan, west side
- Clamp mounted dynamic strain gauge at the bottom flange of midspan floor beam;
- Drilled dynamic strain gauge at the bottom chord west side near midspan

For main bridge

- Uni-axial accelerometer measuring longitudinal motions at north end of north anchor span (pier 14) and at south end of south anchor span (pier 17)
- Two uni-axial accelerometers measuring vertical motion (east and west sides top chord) and one measuring transverse motions (west side top chord) repeated at five locations on the main bridge (in North and south anchor spans; north and south cantilever spans and in suspended span)
- Additional uni-axial accelerometer measuring transverse motions at west side bottom chord at two locations; North and South cantilever spans.
- Strain gauge similar to approaches (on floor beam and bottom chord) at five locations similar to accelerometer locations on main bridge (10 total)
- Wind direction and wind speed sensor located on west side, bottom chord near midspan of suspended span.

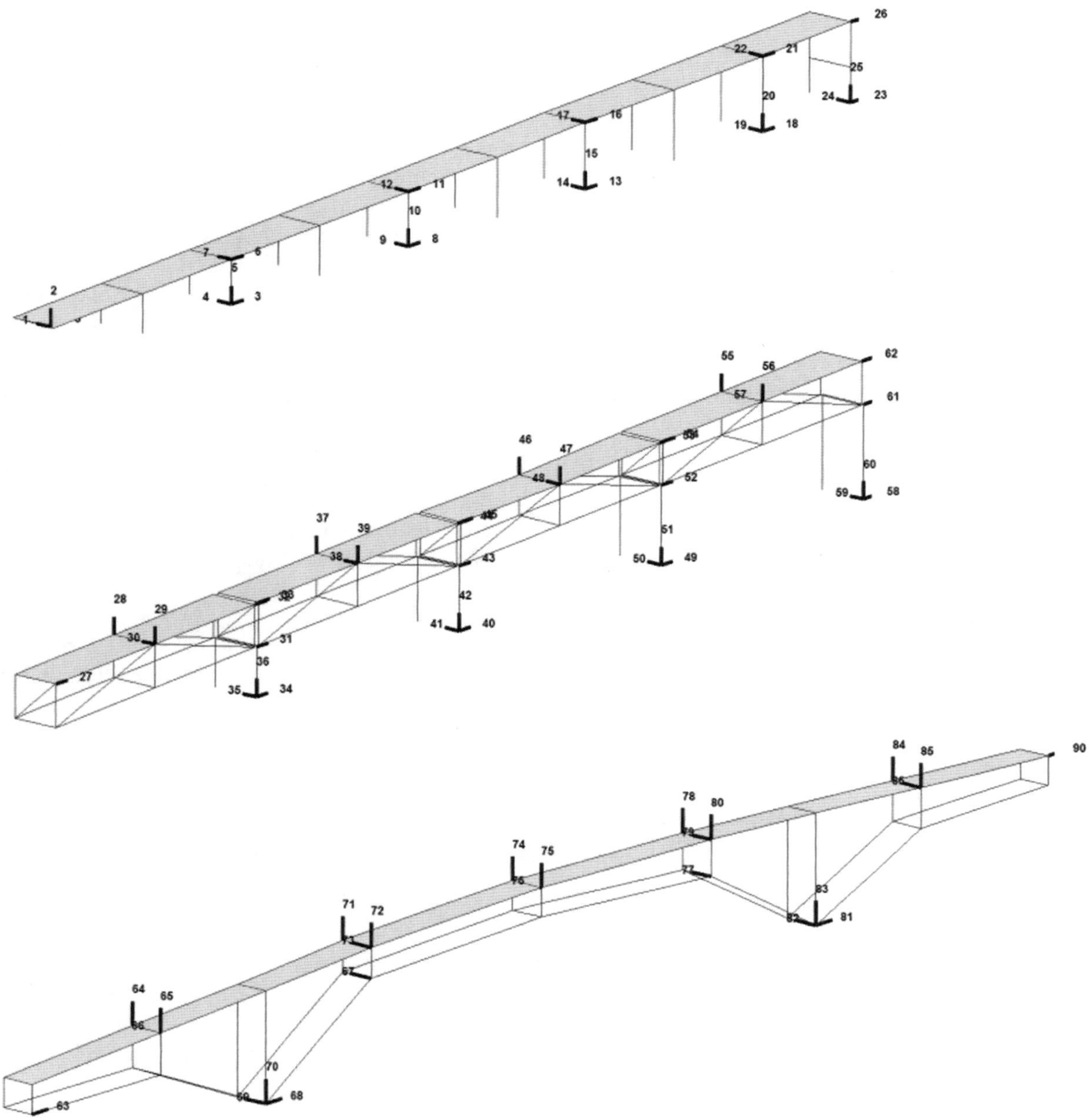

Fig. 6.3 Location of accelerometers on idealized model of bridge; Viaduct *(top)*, approaches *(middle)* and main span *(bottom)*

Additional ground sensors:

- Tri-axial sensor mounted on pier foundation at piers 10, 11, 12, 13, 14, 15, 16 and on south abutment (17)
- Two tri-axial free-field sensors: North end on concrete pad between Piers 10 and 11; South end in concrete vault on East side of HW 1 south of IMSNB monument.
- Downhole triaxial accelerometer 12 m deep between Piers 10 and 11 (underneath North free-field sensor)

Additional details

- Two temperature sensors; at Piers 10 and 16 near top of pier
- Central DAQ (data recorder) located at foundation level of Pier 10
- Back-up power system consisting of battery UPS (12 h) and propane generator (20 day) located at foundation level of Pier 11 (Fig. 6.3)

6.4 Comparison of Traffic Load to Modal frequency

One set of monitored ambient vibration data (17 days in length) is collected from IMSNC Bridge starting from August 10th, 2012 at 18:00 p.m. to August 27th, 2012 at 15:00 p.m. Collected data is analysed with an algorithm developed in MATLAB [14]. The algorithm includes calculation of Fourier Amplitude Spectrum (FAS), Root Mean Square (RMS) acceleration, and mean value of each channel, the identification of modal frequencies, and time variations of calculated parameters. Standard signal processing tools are applied to each acceleration channel: baseline correction and band-pass filtering. The corner frequencies of band-pass filter are selected as 0.05 and 20 Hz for high-pass and low-pass, respectively. Results presented in this study focus on only the midspan of the bridge, which incorporates the bents between 14 and 17.

Figure 6.4 depicts the time variation of FAS for channel 74, which is located at mainspan of the bridge and is measuring the vibrations in vertical direction. The first modal frequency is at 0.8 Hz, which is the first vertical mode of the main span of the IMSNC Bridge. The amplitudes increase in daytime and reach its peak values at around 12:00 p.m. every day due to increase in traffic load on the bridge. Conversely, the amplitudes of the first vertical mode decrease in night due to decrease in traffic load. Such increasing and decreasing in amplitudes is observed for each day during 17 days of recording. Because there is less traffic on the bridge in weekends, the amplitude of the first vertical mode does not experience the same level of amplitude as it does in weekdays. The second peak, as seen from Fig. 6.5, is at around 1.167 Hz, which is the first torsional mode of the mainspan bridge. Table 6.1 lists the first four vertical and the first three torsional modal frequencies of the mainspan of IMSNC bridge. Similar amplitude fluctuation behaviour is observed for the first torsional mode of the IMSNC bridge listed in Table 6.1: increase in daytime and decrease at night. The paper focuses on the first vertical mode and the first torsional mode of the mainspan of ISMNC bridge (Fig. 6.6).

The traffic load on the bridge varies during daytime and night, generally adding to the mass of the structure during the day. The vertical amplitudes of the vibration in the midspan of the bridge were observed to have a direct correlation with the traffic load on the bridge: increasing in daytime and decreasing in night. The scale on the right side of the Fig. 6.7 is the RMS value of the vertical vibrations recorded by channel #74, which is located in the midspan of the bridge Channel #75, is the vertical sensor on the opposite side of the bridge. The figure clearly shows the increase in vibration in daytime and decrease in night due to traffic load on the bridge. RMS values, therefore, can be interpreted as an indicator of the traffic load on the bridge.

The scale on the left side of Fig. 6.7 is the modal frequency. The modal frequency varies between 0.75 and 0.82 Hz: decreasing in daytime and increasing at night. Since the RMS value increases in day and decreases at night, the fluctuation in modal frequency coincides with the traffic load (RMS values). Similar behaviour is observed for the first torsional mode of the mainsapan of the bridge where the first torsional mode varies between 1.15 and 1.175 Hz. It could then be further inferred that the change in mass during the day with additional traffic load is what affects the reduction in modal frequency.

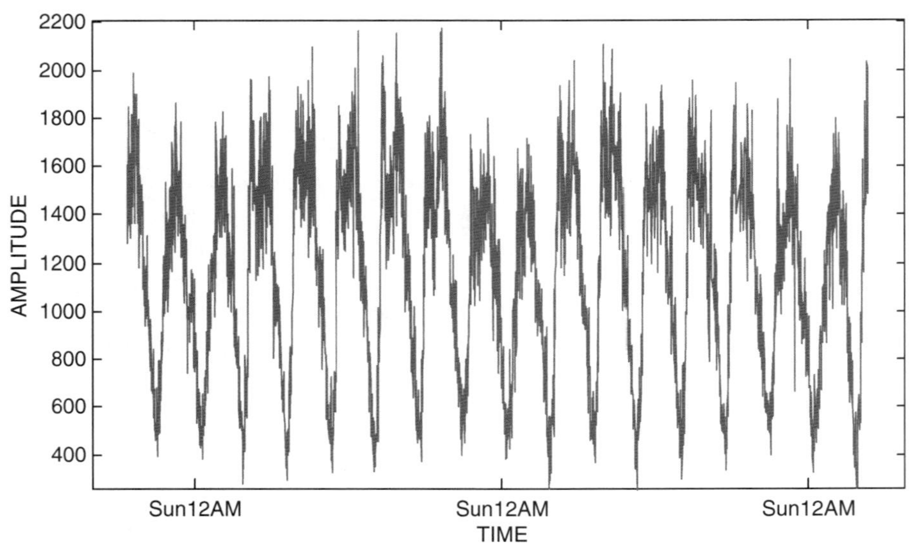

Fig. 6.4 Time variation of Fourier Amplitude Spectrum of the first vertical mode at 0.8 Hz of channel #74, which is located in the mid span of main bridge and is recording in vertical direction

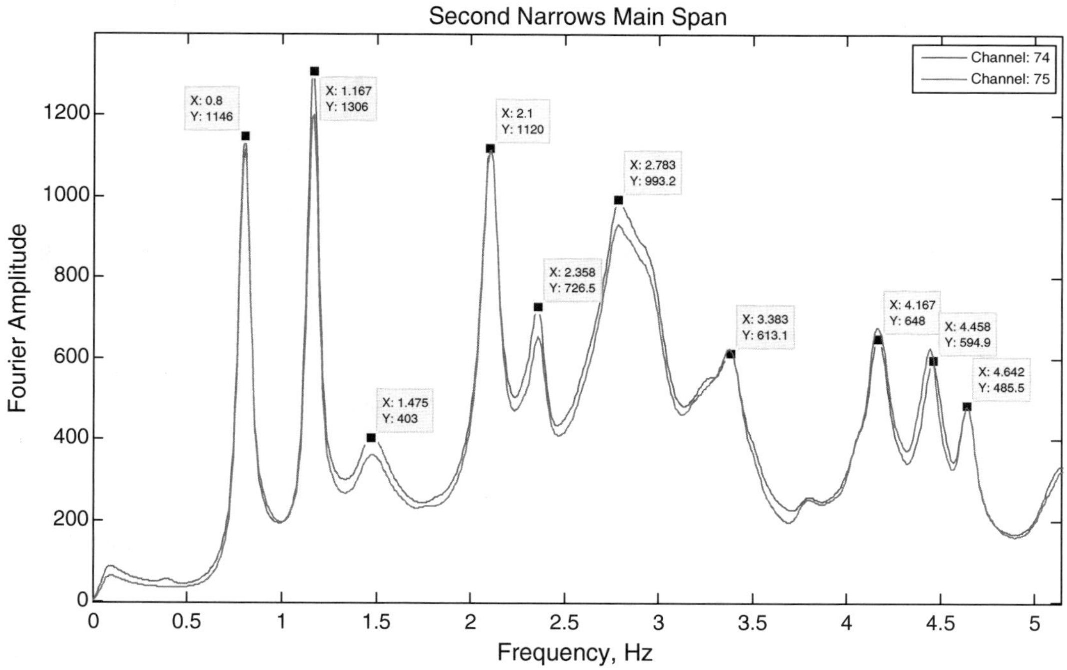

Fig. 6.5 The average of FASs – for channels 74 and 75 – over 17-day clearly indicates the vertical and torsional modal frequencies of the main span. Averaging filters out all noise components from entire frequency band

Table 6.1 Vertical and torsional modal frequencies of the IMSNC Bridge (mainspan only)

Mode	Freq. [Hz]	Descr.
1	0.80	1st vertical
2	1.167	1st torsion
3	1.475	2nd vertical
4	2.10	3rd vertical
5	2.358	2nd torsion
6	2.783	4th vertical
7	3.383	3rd torsion

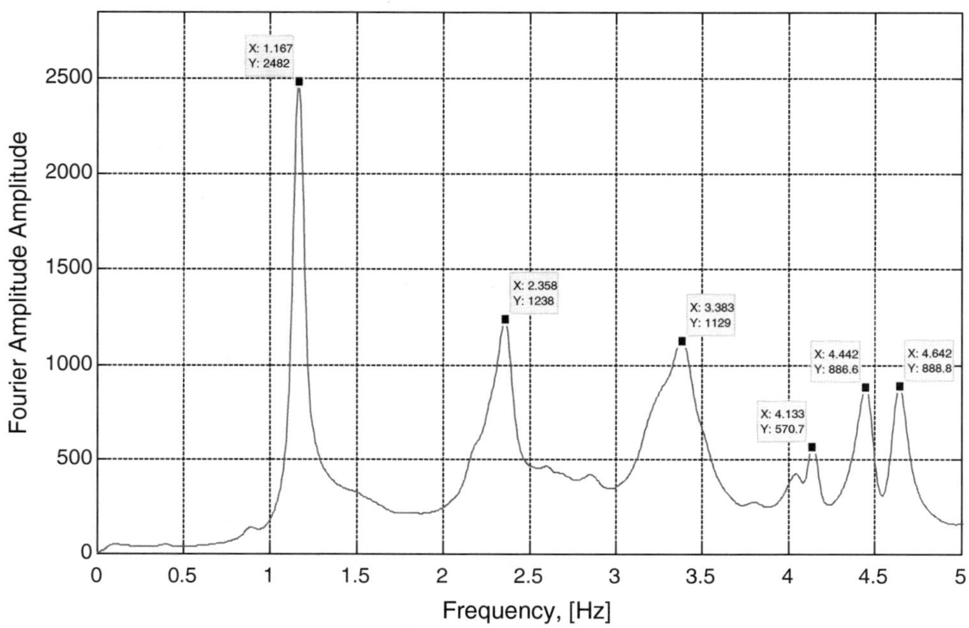

Fig. 6.6 Average of Fourier Amplitude Spectrum of torsional modes of the mainspan (obtained by the difference of channel #74 and 75)

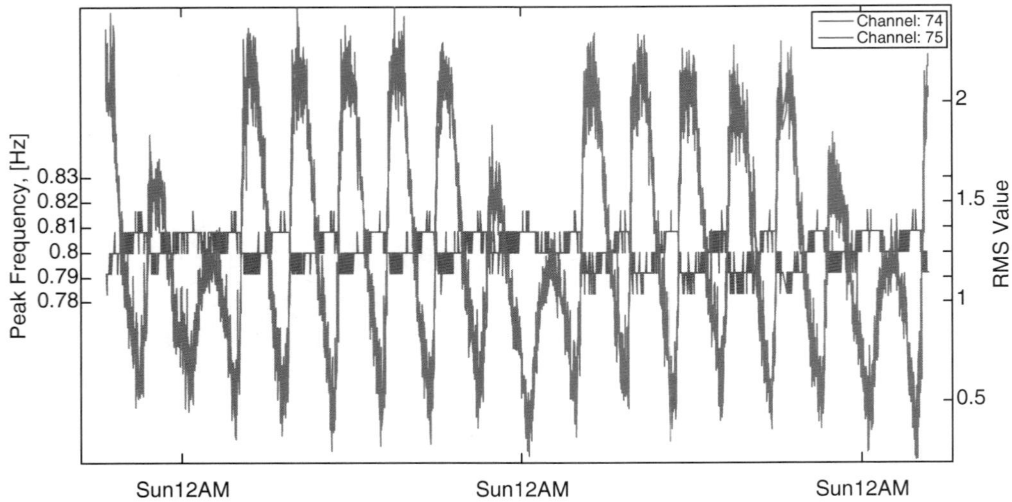

Fig. 6.7 The time variation of the first vertical mode of the main span (*green and blue lines*) and the RMS values of the recorded ambient vibration (*red-line*) data

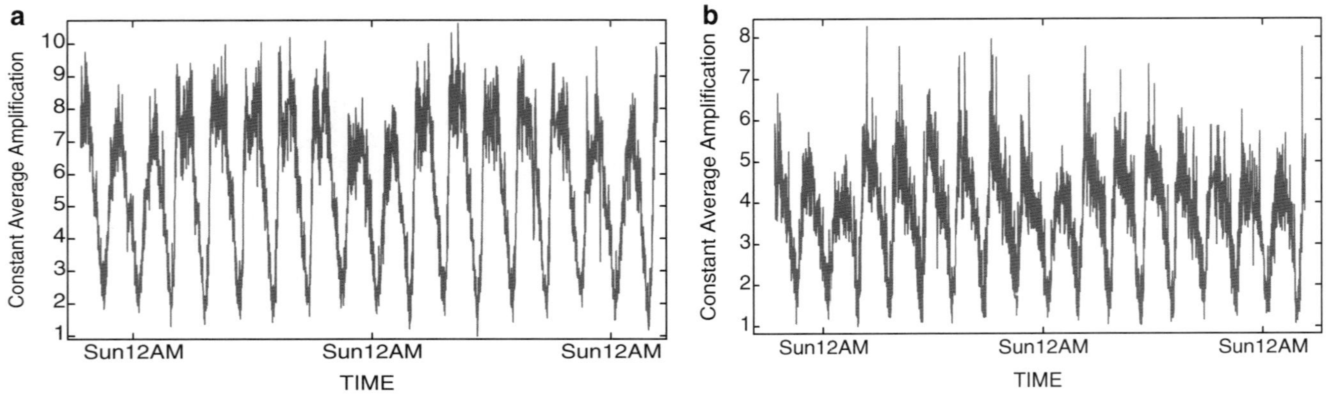

Fig. 6.8 The average amplification factor for: (**a**) the first vertical mode, (**b**) the first torsional mode of the mainspan of the IMSNC bridge

A recent work at The University of Nevada at Reno studied the effect of traffic on a bridge during a seismic event [15]. It was found that there is a significant improvement on the bridge performance during small amplitude motions due to the presence of vehicle loads. This may be due to increased damping in the vehicle-structure system. Further study to investigate any relationship in changes in damping and modal frequency with traffic load is warranted.

Such fluctuation in the RMS values does not only cause shift in modal frequency, but also causes amplification in the modal response. Figure 6.8 shows the time variation of the amplification of first vertical mode of the bridge. This amplification, C_O, varies for each modal frequency and is calculated for 2 min of moving windows. The amplification is calculated as a constant value for each moving window. The value of the constant is typically determined by taking the average of the standard FASs over a selected frequency band (f_i and f_n) as given in (6.1)

$$C_0 = \frac{\sum_{f=fi}^{fn} |X(f)||Y(f)|}{\sum_{f=fi}^{fn} X^2(f)} \tag{6.1}$$

where $X(f)$ and $Y(f)$ are the Fourier amplitude spectrum of reference state and current state, respectively. The amplification is calculated over a narrow frequency band, f_i and f_n, which contains the modal response for which the amplification is calculated for. Derivation of constant amplification, C_O, is based on least-square and more information about it can be found in [16]. The reference state is taken as the state on Wednesday August 17th at 12:10 a.m. where the constant average amplification is calculated as 1 for the first vertical mode frequency as shown in Fig. 6.8. Amplification of the first vertical mode fluctuates between 1 and 10.2 with respect to reference state. The traffic load on the bridge amplifies the response of the first vertical mode seven and ten times during daytime weekdays and daytime weekends, respectively. The first torsional mode of the structure amplifies 5.5 and 4 times in daytime weekdays and weekends, respectively.

6.5 Comparison of Temperature to Modal Frequency

A plot of the changes in temperature over the period of 17 days is shown in Fig. 6.9. The temperature varies between 9.4 °C and 25.7 °C, with maximum during the day and minimum at night as expected; however the maximum and minimum values are not consistent from a day to day basis, as was seen with the RMS and modal frequencies. Therefore it is not possible to conclude that the temperature effect is directly affecting the change in modal frequency. Further analysis over a longer period of time is will be performed to have a better understanding about the relation between temperature and modal properties of the IMSNC bridge.

6.6 Summary and Conclusions

This paper presented a study on the monitored change in fundamental frequency for the IMSNC in Vancouver BC. Examining 17 days of data, the fundamental frequency was shown to vary between 0.75 and 0.82 Hz in a roughly sinusoidal pattern with the highest frequency occurring at night and the lowest occurring during the day. The change in frequency was compared to changes in monitored temperature and in RMS acceleration of the vertical sensors at midspan of the bridge. From the results, it was found that:

- The fundamental frequency of the bridge (vertical mainspan) varied between 0.75 and 0.82 Hz in a daily pattern, with maximum (approximately) at midnight and minimum (approximately) at noon. Similarly, the first torsional mode of the midspan of the bridge varied between 1.15 and 1.175 Hz with maximum at midnight and minimum at noon.
- The RMS amplitude of the vertical acceleration at midspan varied between 0.22 g at midnight and 2.48 g at noon, which corresponds well with general traffic on the bridge having the highest levels during the day with reduced traffic at night. Further, upon examination of a weekly pattern, it is seen that the fluctuation is regular during the 5 weekdays, with less amplitude on Saturday, and the least amplitude on Sunday.
- Temperature is seen to vary normally with peaks during the day and lows at night. The variation over the 17 day period is between 9.4 °C and 25.7 °C.
- The traffic load on the bridge amplifies the response of the first vertical mode seven and ten times during daytime weekdays and daytime weekends, respectively. The first torsional mode of the structure amplifies 5.5 and 4 times in daytime weekdays and weekends, respectively.

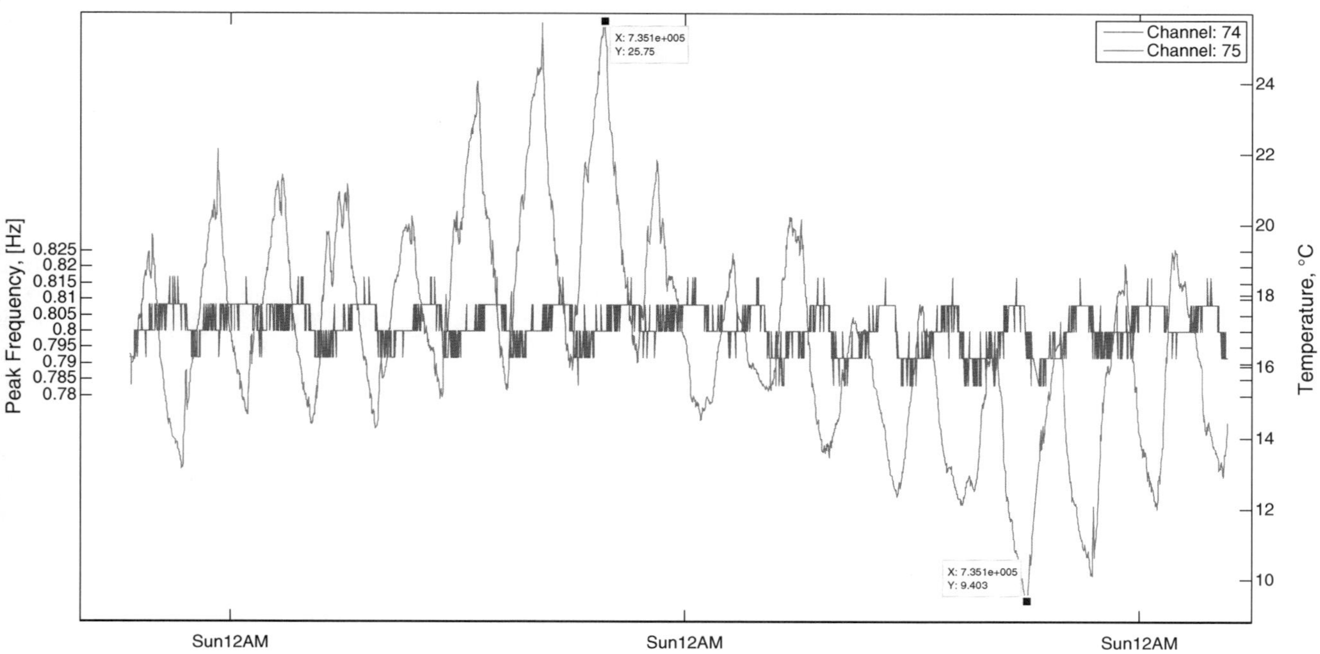

Fig. 6.9 The time variation of the first vertical mode of the main span *(green and blue lines)* and the measured temperature values

Since the observed changes in frequency and amplitude are more regular than the changes in temperature, it could be said that the change in frequency is due to the effect of traffic on the bridge. This could be due to the direct effect of the increased mass on the bridge; it is seen that the lowest frequency is during the day (and highest at night) which corresponds with the concept of the greater mass during in the day lowering the fundamental frequency.

To examine this further, a more detailed study will be required to look at the effect of change in traffic throughout the day. For example in the two rush hour periods where the bridge is heavily loaded with slow moving traffic, and then the midday period where the bridge is heavily loaded with faster moving traffic. In addition the phenomenon will be studied using a longer set of data, such as sets of 2 weeks of data each month for 6 months.

References

1. Khanukhov KM, Polyak VS, Avtandilyan GI, Vizir PL (1986) Dynamic elasticity modulus for low-carbon steel in the climactic temperature range (trans: Problemy Prochnosti), vol 7. Central Scientific-Research Institute of Designing Steel Structures, Moscow, pp 55–58
2. Sabeur H, Colina H, Bejjani M (2007) Elastic strain Young's modulus variation during uniform heating of concrete. Mag Concre Res 59 (8):559–566
3. Sohn H (2003) Effects of environmental and operational variability on structural health monitoring. Philos Trans R Soc A365(1851):539–560
4. Nurdan A, Kaya Y (2012) Vibration characteristics of a suspension bridge under traffic and no traffic conditions. Earthq Eng Struct Dyn 41 (12):1717–1723(7), 10 Oct 2012
5. Cornwell P, Farrar CR, Doebling SW, Sohn H (1999) Environmental variability of modal properties. Exp Tech 23(6):45–48
6. He X (2008) Vibration-based damage identification and health monitoring of civil structures. Ph.D. thesis, Department of Structural Engineering, University of California, San Diego
7. Safak E (2005b) Detection of seismic damage in structures from continuous vibration records (invited paper). In: Proceedings of 9th international conference on structural safety and reliability (ICOSSAR), Rome, 19–23 June 2005
8. Clinton JF, Bradford SC, Heaton TH, Favela J (2004) The observed wandering of natural frequencies in a structure. Bull Seismol Soc Am (in press)
9. Ni YQ et al (2005) Correlation modal properties with temperature using long-term monitoring data and support vector machine technique. Eng Struct 27:1762–1773, Elsevier, 25 July 2005
10. Hoon S et al (1999) An experimental study of temperature effect on modal parameters of the Alamosa Canyon bridge. Earthq Eng Struct Dyn 28:879–897
11. Yong X (2006) Long term vibration monitoring of an RC slab: temperature and humidity effect. Eng Struct 28:441–452
12. Ventura C, Kaya Y (2012) Seismic structural health monitoring of bridges in British Columbia, Canada. In: Proceedings of the 15 world conference of erathquake engineering, Lisbon
13. Turek M, Ventura C, Dascotte E (2010) Model updating of the ironworkers memorial second narrows bridge, Vancouver, Canada. In: Proceedings of the IMAC-XXVIII, Jacksonville, 1–4 Feb 2010
14. MATLAB R (2001b) The MathWorks Inc., Natick
15. Wibowo H, Sandford DM, Buckle I, Sanders D (2012) Evaluation of vehicle bridge interaction during earthquake. In: Proceedings of 15WCEE Lisbon, Paper 1560, Portugal
16. Safak E (1997) Models and methods to characterize site amplification from a pair of records. Earthq Spectra EERI 13(1):97–129

Chapter 7
Structural Monitoring and Analysis of Bridges for Emergency Response

Martin Turek, Yavuz Kaya, Carlos Ventura, and Sharlie Huffman

Abstract The British Columbia Ministry of Transportation and the University of British Columbia have implemented a program to instrument key structures to provide confirmation of seismic capacity, assist in focusing retrofit efforts, perform structural health evaluations and provide rapid damage assessment of those structures following a seismic event. The instrumentation system installed at each structure will automatically process and upload data to a central server via the Internet. The alert systems and public-access web pages can display real time seismic data from the structures and from the BC Strong Motion Network to provide input for assessments by the Ministry of non-instrumented bridges. These systems may also provide other agencies, emergency responders and engineers with situational awareness.

Keywords Instrumentation • Real-time monitoring • Emergency response • Damage detection • Modal identification

7.1 Introduction

The MoT has been instrumenting structures in collaboration with the Earthquake Engineering Research Facility (EERF) at the University of British Columbia (UBC) since the late 1990s. The west coast of BC lies in Canada's highest seismic zone, a as a result, the primary purpose of these original systems was to capture the ground motion input in the event of an earthquake. More recently, the instrumentation has been expanded to incorporate Structural Health Monitoring (SHM). Two design- build bridges have included instrumentation; one existing bridge has also been instrumented, and up to eight more bridges will be added to system by the end of 2014.

In addition to the structural monitoring, the Geologic Survey of Canada (GSC) through the Pacific Geosciences Centre (PGC) maintains the Provincial Strong Motion Network (SMN) comprised of over 130 ground monitoring stations. Over the last several years the MoT has been working with the PGC expanding the number of stations in the network. Building on these collaborations, MoT and UBC embarked on a program called the British Columbia Smart Infrastructure Monitoring System (BCSIMS). The system integrates data from the instrumented structures and strong motion network, organizes and processes the information in an efficient manner, to deliver that information to the appropriate parties.

The Goals of the System are to: (1) Provide a real-time seismic structural response system to enable rapid deployment and prioritized inspections of the Ministry's structures; and (2) Develop and implement a health monitoring program to address the need for safe and cost-effective operation of structures in BC.

M. Turek • S. Huffman
British Columbia Ministry of Transportation and Infrastructure, Victoria, Canada

Y. Kaya (✉) • C. Ventura
University of British Columbia, Vancouver, Canada
e-mail: kayaya@mail.ubc.ca; ventura@civil.ubc.ca

A. Cunha (ed.), *Topics in Dynamics of Bridges, Volume 3: Proceedings of the 31st IMAC, A Conference on Structural Dynamics, 2013*, Conference Proceedings of the Society for Experimental Mechanics Series 38,
DOI 10.1007/978-1-4614-6519-5_7, © The Society for Experimental Mechanics, Inc. 2013

7.2 BCSIMS System Architecture

Figure 7.1 presents an overview of the BCSIMS system architecture. The system combines components of hardware and data acquisition, data storage and processing, and network communication. As shown in the Figure, structural data is acquired on a local level, using as an example the Second Narrows Bridge (a total of four generic structural stations are shown, but an indefinite number of stations can be added to the system). The raw data from the local DAQ is sent to a local database, which contains all of the same fields as the global database. Event triggers are set at the structural station in the hardware; a trigger initiates a recording that is placed in the local database. This initiates the SIMS2 local analysis module; the results of the analysis are placed in the local database, and both raw data and results/parameters sync to the global database. Upon receipt of new data/results in the global database, the SIMS3 advanced analysis PC will initialize a report generator. Reports will be sent to the webpage and to selected user email accounts via HTML.

In order to streamline the data transmission process, UBC has developed its own data archiving standards and protocols. The advantage of this approach is that it helps achieve consistency and platform-neutrality across all hardware platforms thereby simplifying the downstream processing. In addition to being able to use the most suitable hardware for a specific bridge (e.g. for technical performance or cost effectiveness), it also offers flexibility in replacing (e.g. defective) sensors in the future.

The implementation of the data transfer protocol has been through one of two ways. A Windows COM interface is created that runs on the site PC next to the supplier software. This interface converts the data into the SIMS format and uploads it directly to the local database for use in the BCSIMS system. The second approach is through an application called the 'Virtual Data Acquisition System', VDAS, which will also run on the site PC. Since typically each hardware supplier has its own data format, VDAS is designed in such a way that it can read different data formats, converts them to SIMS format, and uploads them into a ring buffer in local database.

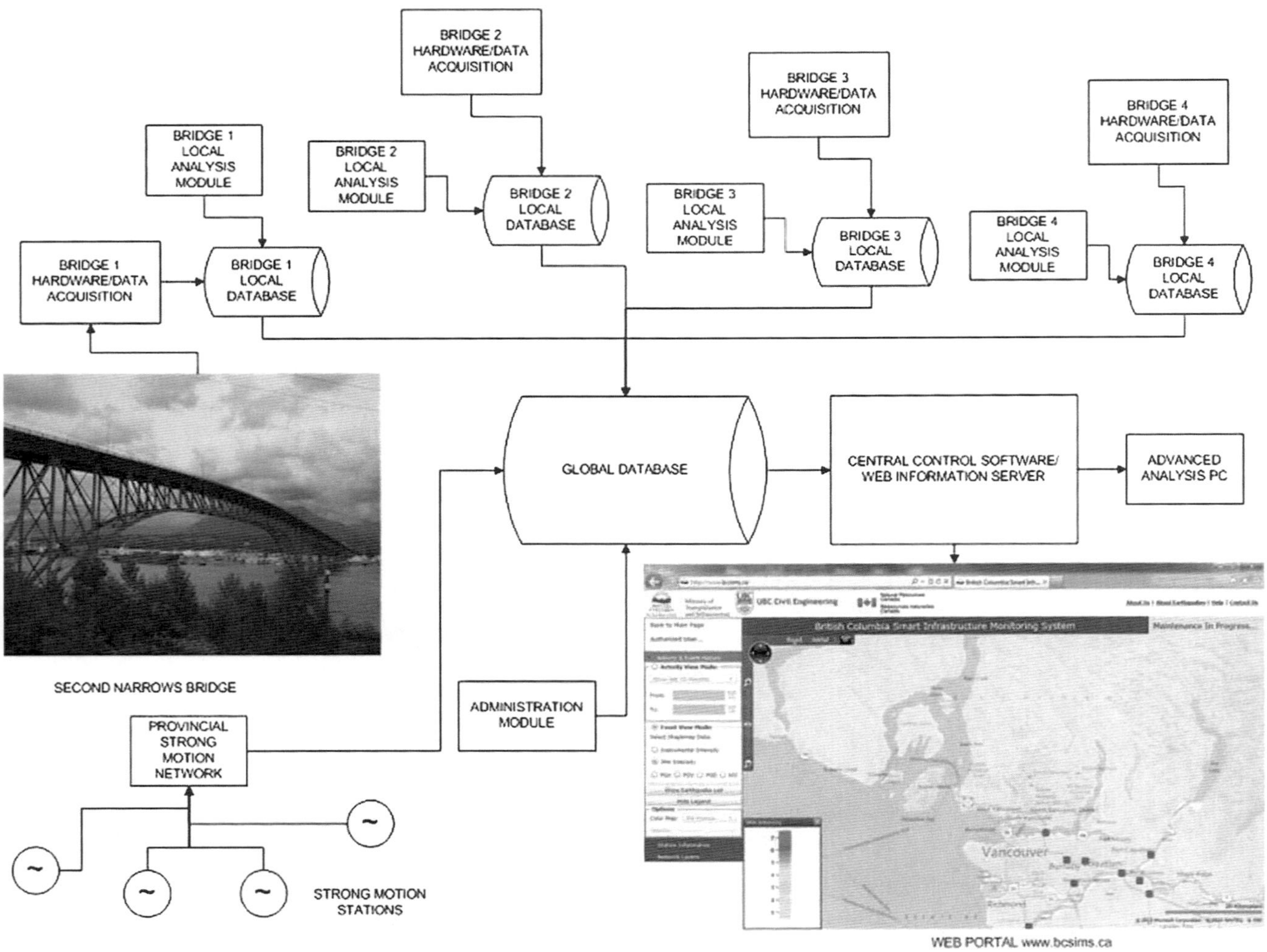

Fig. 7.1 BCSIMS system architecture

7.3 Web Interface

The www.bcsims.ca website is the gateway for user interaction and operational management. There are two view modes – public and restricted, which dictate the nature and amount of information accessible on the webpages. The public view consists of a shakemap intended for general public consumption and the restricted view contains further information for advanced users, such as processed and/or raw downloadable data and unpublished results. Figure 7.2 shows a screen shot of the website homepage. The circles represent the strong motion network stations, and the squares are structural stations.

The structure stations and strong motion sensors of the IANet are displayed as icons on a digital geomap. The interactive map allows zooming in/out and focusing on a particular station. Additional metadata for the structures such as location information and live links to webcams are also provided. Lists of recent events and recent seismic activities are provided from which the user can access published information for the corresponding events and activities.

There are two main view modes in the main page of the bcsims.ca from which the user can extract information: Station Activity & Event view mode and Earthquake Events view mode. Station Activity & Event mode displays activities recorded for each IANet station within a user selectable time period. Recorded activities are indicated by color and circle. Size of the circle indicate the maximum measured amplitude of the record (PGA), where as color depicts the kSI (Katayama Spectral Intensity) value. By clicking on any of the IANet station on the geomap, the user can access all the activities recorded by that station, and download raw or process data such as velocity, displacement, Fourier amplitude spectrum, and response spectrum. Earthquake Event View mode, on the other hand, displays recent earthquakes that has happened in BC with its epicenter located on geomap. The epicenter of an earthquake is revealed by a colored-star where the size of star indicates the moment magnitude of the earthquake. The user has the option to display several different shake maps such as Instrumental Intensity [1], Katayama Spectra Intensity [2], and Japan Meteorological Agency Intensity [3]. Other options include the, PGA, PGV and PGD values. The default option is the PGA shake map.

In addition to the homepage and shakemaps, links from each bridge icon in the map directs to the Structures Information Pages. The idea of the Structures Information Page (SIP) is to provide an overview of the status of the structure and more detailed results of the various structural assessments carried out by the system. The SIP is divided into several tabbed sections. The tabs include a summary view, analysis view, data view and structure view. The exact configuration of the tabs

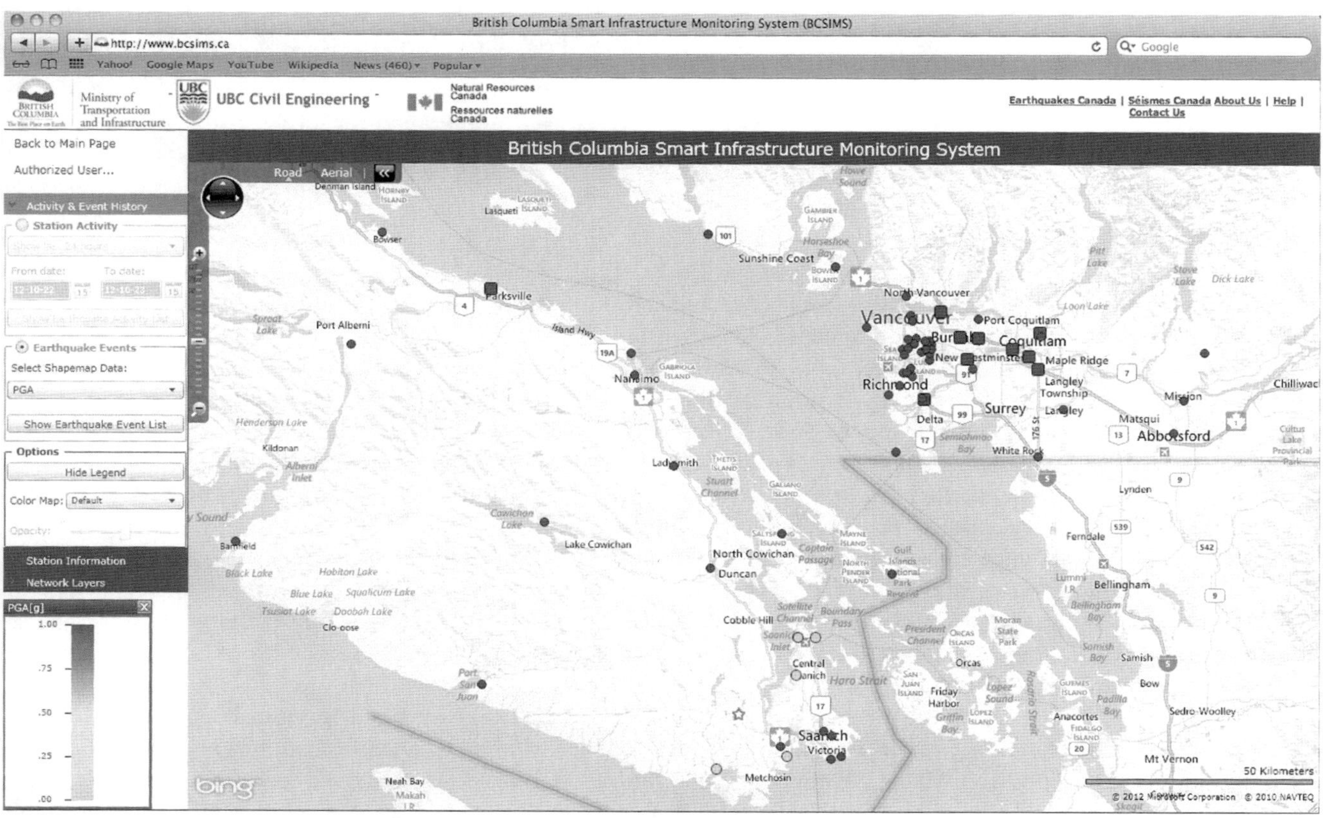

Fig. 7.2 Screenshot of the BCSIMS homepage

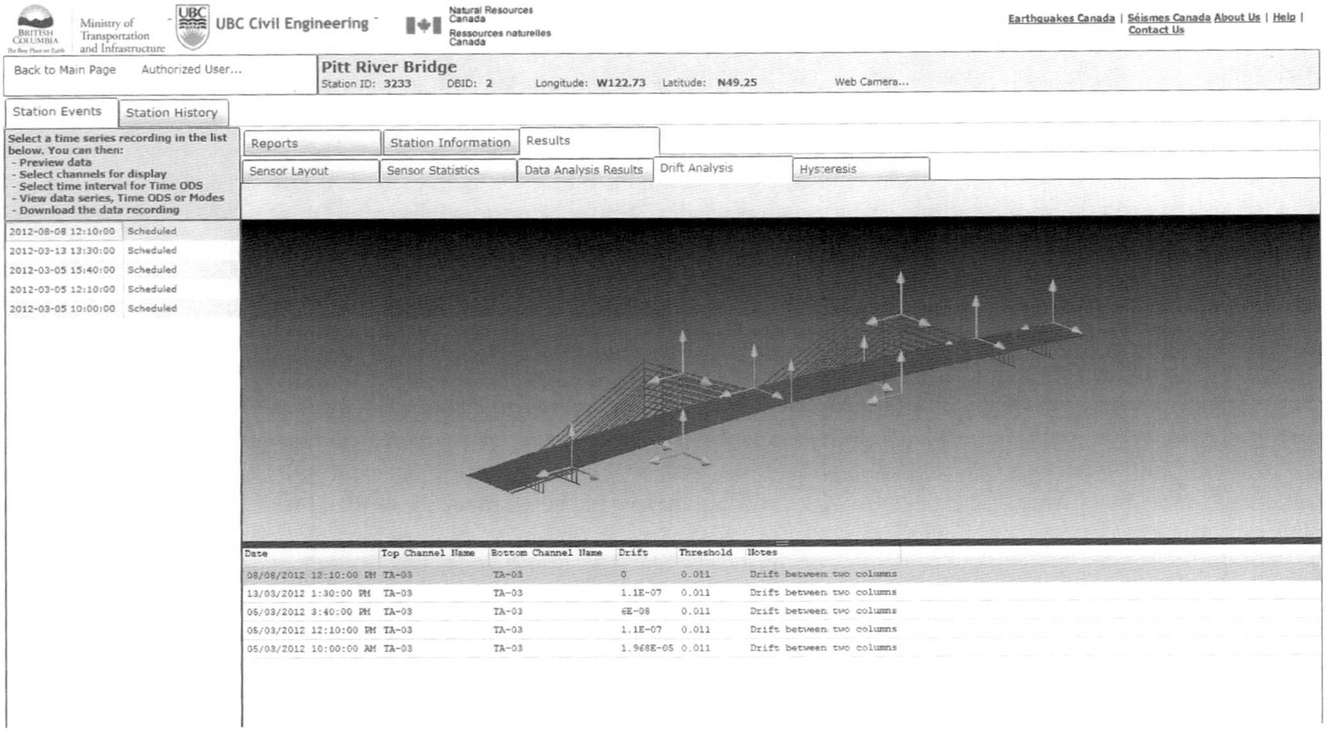

Fig. 7.3 Structure information page 3D model view

is being finalized and may differ from this list; however the content is generally the same. The concept is to separate into information relating to events, data, analysis and static data such as instrumentation drawings and photos of the site. Figure 7.3 shows the current 'Drift Analysis' (which contains data); a 3D model of the Pitt River Bridge is shown. The model can be used for various analysis results.

7.4 Structural Monitoring

The MoT has been instrumenting bridges and tunnels in collaboration with the University of British Columbia since the late 1990s; four structures were originally instrumented. These include the French Creek Bridge and Portage Creek Bridge on Vancouver Island, George Massey Tunnel (Hwy 99) and Queensborough Bridge (Hwy 91A) both South of Vancouver. Work is underway to bring these legacy systems online into the BCSIMS network. These bridges typically featured strong-motion type accelerometers, in downhole, on footings and on the structure. There are also some downhole piezometers (measuring pore pressure) and strain gauges (measuring strain on glass-fiber column wraps).

In 2008 the new W.R. Bennett Bridge opened on Highway 97 near Kelowna, BC. The partially floating bridge was instrumented with a 12-channel accelerometer system. In 2009, the new Pitt River Bridge opened on Highway 7 near Maple Ridge, BC. The cable-stayed bridge was instrumented with a 46-channel system including accelerometers and wind sensor. In 2011, instrumentation was installed on the 50-year old Ironworkers Memorial Second Narrows Crossing Bridge on Highway 1, between Vancouver and North Vancouver. The system features 122 channels of accelerometers, free-fields, downhole accelerometer, strain gauges, and temperature and wind sensors. Pitt River and Second Narrows were the first two bridges to be implemented into BCSIMS.

Upcoming instrumentation includes the new Port Mann Bridge (Fig. 7.4) on Highway 1 between Coquitlam and Surrey, BC. The 10-lane cable-stayed bridge will be instrumented with cable and structure accelerometers, wind, temperature and humidity sensors, as well as ground and downhole stations. This is a part of a major infrastructure project that will also include instrumentation on three underpass bridges, one 650 m long twin steel girder viaduct type structure, and seven more strong motion network stations. All of these systems will be implemented to the BCSIMS network.

Fig. 7.4 Port mann bridge

7.5 Data Analysis

Data analysis is a core functionality of the BCSIMS system. Acknowledging the high cost and risk associated with continuous data-streaming, founding philosophies of the system is to process data at the site and transmit results in real time only; sending full data only as the connection/bandwidth is available.

As shown in Fig. 7.1, the data analysis takes place at a variety of stages and locations. Data is analyzed at the structure site by the SIMS2 module and at the central site by the SIMS3 PC. Post processing of data includes statistics, ground motion analysis, modal analysis, drift and hysteresis analysis, damage detection, finite element analysis, and finite element model updating. All of these results are stored in the global database either directly by SIMS3, or by the process of first storing in the local database, and synchronizing back to the global database by SIMS2 module.

Statistics are generated at the structure site on all of the data channelsby the SIMS2 module. It is capable of calculating the mean, RMS, max, min, peak, standard deviation, skewness, and kurtosis. Such statistical values help to better understand the structural behavior under different loading conditions, such as seasonal temperature change, daily traffic loads on bridge, etc.

The raw ground motion data is processed and stored in the COSMOS Strong Motion Data Format [4]. This well-known data format creates three sets of files. Volume 1 files contain the raw data converted to physical units; these are typically referred to as the 'uncorrected files'. Volume 2 files contain the products of the processed raw time history data from Volume 1. This includes correction for instrument response and digital filtering. The velocity and displacement are also obtained using numerical integration. Volume 3 files contain all of the spectral products, including the response spectra (absolute acceleration, relative velocity, and relative displacement). Also included is the Fourier amplitude spectrum.

The SIMS2 module performs modal analysis on the acceleration time histories by means of the time domain Stochastic Subspace Identification technique as implemented in the ARTeMIS Extractor software [5]. The output of the analysis provides the identified natural frequencies, mode shapes, and damping ratio for a given dataset. These results are stored in the local/global database and can be viewed on a 3D model of the structure through the web interface. The frequencies are posted on a Control Chart and can be tracked against time, depending on the frequency of scheduled measurements. The identified modal properties are also used in the model updating process.

Functionality for drift analysis is setup in the SIMS2 module; the user specifies drift pairs and the system computes the drift from the integrated displacement values. The displacement values are calculated by double integration of the narrow-band filtered acceleration data. The filter corner frequencies of each drift analysis depend on the first predominant frequency of the part of the structure for which the drift is calculated. The peak displacement values calculated during an event are stored in the database. Any drift value exceeding predefined threshold value will indicate a possible damage in the structure.

The BCSIMS system is set up to use various techniques of damage detection at different locations. In addition to the techniques already implemented in the system, part of the mandate of the BCSIMS scope is to conduct research on new techniques and implementation of other existing techniques. The first of the implemented algorithms is in the SIMS2 module. It is a statistical algorithm [6]. As with many algorithms, the method observes damage as changes to modal parameters. Consequently, the first requirement of the method is to obtain a series of data sets from the same structure, to obtain a baseline model from which to observe potential changes.

One desirable feature for this algorithm is its ability to detect damage even in the presence of noise, and common environmental effects such as temperature changes. An additional advantage is with regards to the concept of the sub-structuring or clustering of the model for the analysis. This is important for the speed of the algorithm, and for the accuracy of the damage identification. In addition, parts of the structure, which are not expected to be damaged, can be removed from the analysis to improve on speed and accuracy.

A second set of algorithms has been implemented to run in an offline mode through the SIMS3 machine. Five of those can be considered as mode shape based methods. These methods were compiled and used in a well-known study by [7]. These algorithms are:

1. Damage index method – [8]
2. Mode shape curvature method – [9]
3. Change in flexibility method – [10]
4. Change in uniform load surface curvature – [11]
5. Change in stiffness method – [12]

The primary advantage of these methods is the relative ease of formulation, speed of use and simplicity of output results. However, the disadvantage is in the requirement for measurement of the mode shapes in more detail on the structure. This can present a problem in many situations when limited instrumentation is available.

The last set of methods is the flexibility-based methods, of which two are implemented also in an offline mode on SIMS3. The flexibility-based methods use the concept of changes in assembled flexibility matrix to identify and locate damage in a structure. The flexibility matrix can be obtained for stochastic (output-only) data by manipulations of the results from time domain system identification. Previous work [13] utilized two variations of stochastic flexibility methods. These methods are:

1. Stochastic Damage Locating Vector (SDLV) Technique – [14], based on the Damage Locating Vector (DLV) Technique – [15, 16].
2. Proportional Flexibility Matrix Technique – [17]

Finite element models of the monitored structure are another important element in the overall data analysis capabilities. Software is setup on the SIMS3 for the analysis; both during initial calibration and during routine and triggered events. The models will be updated using data from the measured structure, and are used for analysis such as:

1. Stress and load
2. Fatigue
3. Damage location and quantification
4. Prognosis (life expectancy)

For most monitoring system cases, a preliminary FEM will be created and an on-site ambient vibration test will be performed. The FEM will then be manually or automatically updated based on the obtained results. The model can then be used to:

1. Design the permanent monitoring system
2. Evaluate potential damage detection methods through simulation
3. Perform real-time analysis
4. Be used for scheduled structural analysis based on updated models

Automated FEM updating is performed through the SIMS3 PC using the FEMTools commercially available software. Once a new set of modes is placed in the database, an automated process on the SIMS3 machine triggers an updating run using the software; the updated modal comparison matrix and parameter changes are placed in the database. The new FE model is placed in a folder labeled with the event ID and time. The new updated model can be used for further analysis. The process requires a preliminary manual updating of the model, usually done with data from a more detailed ambient vibration test. The settings from this 'manual' update are used to create command files for the automated process.

7.6 Summury and Future Work

The British Columbia Ministry of Transportation and the University of British Columbia have embarked on a program called the British Columbia Smart Infrastructure Monitoring System (BCSIMS). The system aims to integrate data from the instrumented structures and the strong motion network, organize and process the information in an efficient manner, and to deliver that information to the appropriate parties.

The Goals of the System are to: (1) Provide a real-time seismic structural response system to enable rapid deployment and prioritized inspections of the Ministry's structures; and (2) Develop and implement a health monitoring program to address the need for safe and cost-effective operation of structures in BC. Currently the system incorporates more than 100 strong motion network stations, five structural stations, and as many as ten more structural stations by the end of 2014.

The system is based on local database and analysis modules, located at every structural monitoring site. A global database, web server and advanced analysis PC are located at the University of British Columbia and act as the heart of the system. Access for most users is via the website www.BCSIMS.ca which allows for viewing of the strong motion network, structural stations, data, results and event reports. BCSIMS features some other capabilities that are discussed in more detail in [18].

The next phase of development of the BCSIMS system will be through the European Union funded ISMS Project. This will feature development and implementation of new damage detection algorithms. It will also feature several upgrades to the existing BCSIMS framework such as more sophisticated graphical interfaces, expansion and revision of the current database functionality and more efficient analysis methods. Currently the system is being tested, and work with BC emergency management teams is underway to begin using the system.

Acknowledgements The Authors would like to Acknowledge Dr. Palle Anderson and Dr. Henrik Vollesen of Structural Vibration Solutions, Aalborg Denmark. They have been the primary developers of much of the software components and have worked tirelessly to push the project forward. Also they would like to acknowledge Dr. Andreas Rosenberger from the Pacific Geosciences Centre for his considerable contributions to the project. The primary funding for this project has been provided by the British Columbia Ministry of Transportation and Infrastructure.

References

1. Wald DJ, Quitoriano V, Heaton TH, Kanamori H (1999) Relationship between peak ground acceleration, peak ground velocity, and modified mercalli intensity in California. Earthq Spectra 15(3):557–564
2. Katayama T, Sato N, Saito K (1998) SI-sensor for the Identification of destuctive ground motion. In: Proceedings of ninth world conference of earthquake engineering, vol VII, Tokyo-Kyoto, pp 667–672
3. Karim KR, Yamazaki F (2002) Correlation of JMA instrumental seismic intensity with strong motion parameters. Earthq Eng Struct Dyn 31:1191–1212
4. COSMOS Strong Motion Programs Boards (2001) Consortium of Organizations for Strong-Motion Observations Systems (COSMOS) Strong Motion Data Format, Version 1.20, 15 Aug 2001
5. Structural Vibration Solution A/S (Copyright 1998–2012) ARTeMIS Extractor Pro, Release 5.3
6. Balmes E, Basseville M, Mevel L, Nasser H, Zhou W (2008) Statistical model-based damage localization: a combined subspace-based and substructuring approach. Struct Control Health Monit 15:857–875
7. Farrar C, Jauregui D (1998) Comparitive study of damage identification algorithms applied to a bridge: 1. Experiment. Smart Mater Struct 7:704–719
8. Stubbs N, Kim JT (1994) Field verification of a nondestructive damage localization and severity estimation algorithm, Texas A&M University report prepared for New Mexico State University
9. Pandey AK, Biswas M, Samman MM (1991) Damage detection from changes in curvature mode shapes. J Sound Vib 145:321–332
10. Pandey AK, Biswas M (1994) Damage detection in structures using changes in flexibility. J Sound Vib 169:3–17
11. Zhang Z, Aktan AE (1995) The damage indices for the constructed facilities. In: Proceedings of 13th international modal analysis conference, Nashville, TN USA, vol 2, pp 1520–1529
12. Zimmerman DC, Kaouk M (1994) Structural damage detection using a minimum rank updates theory. J Vib Acoust 116:222–231
13. Turek M, Ventura CE (2007) Calibrated vibration simulations for design of SHM systems. In: Proceedings of IMAC XXV, paper 249, Orlando
14. Bernal D (2006) Flexibility-based damage localization from stochastic realization results. ASCE J Eng Mech 132(6):651–658
15. Zonta D, Bernal D (2006) Strain-based approaches to damage localization in civil structures. In: Proceedings of IMAC XXIV, paper 197, St. Louis
16. Bernal D (2002) Load vectors for damage localization. J Eng Mech ASCE 128(1):7–14
17. Duan Z, Yan G, Ou J, Spencer BF (2005) Damage localization in ambient vibration by constructing proportional flexibility matrix. J Sound Vib 284:455–466, Dynamic Design Solutions N.V. (Copyright 1994–2012). FEMTools Version 3.5.3
18. Ventura C, Kaya Y (2012) Seismic structural health monitoring of bridges in British Columbia, Canada. In: Proceedings of the 15 world conference of erathquake engineering, Lisbon

Chapter 8
Long-Term Modal Analysis of the New Carquinez Long-Span Suspension Bridge

Y. Zhang and J.P. Lynch

Abstract Recent acceleration in the development of sensing and data acquisition technologies has led to the dramatic reduction in the cost of monitoring solutions while functionality has improved. These trends have made instrumentation of large, complex infrastructure systems for long-term monitoring possible. For many structures, long-term monitoring provides a massive data set upon which the system's modal properties can be empirically identified. Modal properties have been previously used for model updating and to better understand the performance of the instrumented structure. In this study, a low-cost wireless sensing platform has been used to permanently instrument a long-span bridge for long-term monitoring of structural behavior under normal operational loading profiles including exposure to wind, temperature and traffic loads. The Alfred Zampa Memorial Bridge (also termed the New Carquinez Suspension Bridge) located in Vallejo, California was instrumented to monitor bridge accelerations, strains, and displacements. In addition, the wireless monitoring system was used to indirectly monitor thermal and wind loads by measuring temperature and wind characteristics (i.e.,direction and speed) respectively. The study's primary focus is on the autonomous extraction of modal characteristics of the bridge using the output-only stochastic subspace identification (SSI) method. Using extracted modal frequencies, relationships are presented between the modal characteristics of the bridge and environmental parameters such as temperature and traffic activity.

Keywords Structural monitoring • Automated modal analysis • Environmental loading • Subspace system identification • Wireless sensors

8.1 Introduction

Long-term monitoring of large-scale civil infrastructure systems is desirable because direct observation of structural responses in normal and extreme load environments can advance the field's understanding of structural behavior. For this reason, many instrumentation projects are underway in the United States. The most successful long-term monitoring project is the California Strong Ground Motion Instrumentation Program (CSMIP) administered by the State of California's Department of Conservation [1]. This program has installed a state-wide array of accelerometers to monitor ground motions and structural responses during earthquakes. In California, many of the long-span bridges including the Alfred Memorial Bridge has a dense array of force balanced accelerometers installed by CSMIP to monitor the bridge during moderate and strong ground motions. While many of these instrumentation programs have been motivated by a desire to understand how bridges behave under extreme loads, recent interest in long-term monitoring has been further motivated by a desire to assess bridge health conditions from monitoring data. This interest in structural health monitoring (SHM) has been triggered by the dramatic reductions in cost of traditional sensing solutions while non-traditional sensing technologies (e.g., wireless sensors) have matured to now offer monitoring conveniences not previously available.

Y. Zhang (✉) • J.P. Lynch
Department of Civil and Environmental Engineering, University of Michigan, 2350 Hayward St,
Ann Arbor MI 48109, US
e-mail: yilan@umich.edu

A. Cunha (ed.), *Topics in Dynamics of Bridges, Volume 3: Proceedings of the 31st IMAC, A Conference on Structural Dynamics, 2013*, Conference Proceedings of the Society for Experimental Mechanics Series 38,
DOI 10.1007/978-1-4614-6519-5_8, © The Society for Experimental Mechanics, Inc. 2013

A number of long-term bridge performance studies have been undertaken on heavily instrumented bridges. In many of these studies, the behavior of the bridges over long periods of study has been described using modal parameters (i.e., modal frequency, modal damping, and/or mode shapes). Modal analysis is well suited for these studies because modal parameters represent the global characteristics of the structure and, in theory, should remain fixed for an invariant system. However, the subtle variations in a structural system's behavior due to environment and time can be efficiently described using modal parameters. Modal properties extracted are also used to calibrate the finite element models of the bridges instrumented. Updated finite element models are essential tools that are used by engineers to estimate the health of a structure after extreme loading events such as strong ground motions. Some examples of long-term bridge performance studies that adopted modal parameters are the long-term structural monitoring systems on the Tamar Bridge in the UK [2] and the Ting Kau Bridge in Hong Kong [3]. The Tamar Bridge is a large-scale suspension bridge that is instrumented with three comprehensive structural monitoring systems. Principle component analysis and mathematical models have been applied to study relationships between modal frequencies and changes in the environment and traffic. The Ting Kau Bridge is a cable-stayed bridge located in Hong Kong and is permanently instrumented with 67 channels of acceleration data and 83 channels of temperature data. Support vector machine techniques have been used to quantify the relationship between modal frequency and temperature.

In this paper, long-term modal analysis is performed on the Alfred Zampa Memorial Bridge (also termed the New Carquinez Suspension Bridge). This bridge has previously been instrumented with a permanent wireless monitoring system to record the behavior of the bridge under its normal operating loads. A total of 124 sensing channels have been installed in the bridge including accelerometers to measure deck and tower accelerations, strain-gages to measure deck flexural responses, and potentiometers to measure deck displacements at the towers. In addition, temperature and wind loads are assessed using thermometers, wind vanes and anemometers. In this study, the accelerometers installed along the bridge deck are primarily used to estimate the modal properties of the bridge using output-only system identification methods. Autonomously extracted modal parameters are analyzed to identify trends in modal parameters due to changes in temperature and time of the day (which will correlate to traffic loading). The paper begins with a description of the New Carquinez Suspension Bridge long-term monitoring study followed by a detailed description of the system identification methods used to extract modal parameters. Next, the results of the study are presented with the paper concluding with a summary of the study's findings.

8.2 New Carquinez Bridge Long-Term Monitoring Study

The New Carquinez Suspension Bridge (Fig. 8.1) is a major suspension bridge built in 2003 over the Carquinez Straights and connects Vallejo with Crockett, CA. The bridge carries four lanes of westbound I-80 traffic. The bridge is constructed from a steel orthotropic box girder supported by two major suspension cables and two reinforced concrete towers (referred to as the south and north towers). The total length of the bridge is over 1,000 m (3,500 ft) and the main span between towers is over 700 m (2,300 ft). A long-term wireless structural monitoring system was installed starting in the summer of 2010 with the system in continuous operation now for over 2 years [4, 5]. The wireless monitoring system installed is based on the use of a low-cost wireless sensor node developed at the University of Michigan termed *Narada* [6]. *Narada* is a wireless data acquisition system specially designed for monitoring civil infrastructure systems where low power consumption (i.e., rechargeable battery operated), high data resolution (i.e., 16-bits or higher), and long communication ranges (i.e., 500 m or longer) are all required system capabilities. On the New Carquinez Suspension Bridge, a total of 33 *Narada* nodes each capable of collecting up to 4 independent channels of data were installed over a 2 year period with various sensing transducers interfaced: 23 tri-axial accelerometers (to measure deck and tower accelerations), 3 string potentialmeters (to measure longitudinal movement between the deck and the towers), 33 battery voltage detectors (to monitor the battery level on each *Narada*), 9 thermistors (to measure the ambient and girder temperatures), 2 wind vanes, 2 anemometers and 6 string gages (to measure deck bending strains). A total of 124 sensing channels have been deployed on the bridge as summarized in Fig. 8.1. The *Narada* nodes (Fig. 8.2a) are all powered by rechargeable batteries that are continuously charged by solar panels installed on the top deck of the bridge (Fig. 8.2b).

The monitoring system is supported by a custom-designed cyber environment that seamlessly integrates the wireless sensors with the Internet where data can be stored in remote database systems and processed using powerful analytical tools. At the bridge, the wireless monitoring system is divided into three sub-networks: one sub-network includes all of the wireless sensors on the girder centered near the north tower (sub-network #1), another sub-network includes all of the wireless sensors on the girder centered near the south tower (sub-network #2), and the third sub-network included the nodes on the top of the two towers (sub-network #3). Each sub-network is managed by a Linux server implemented on an

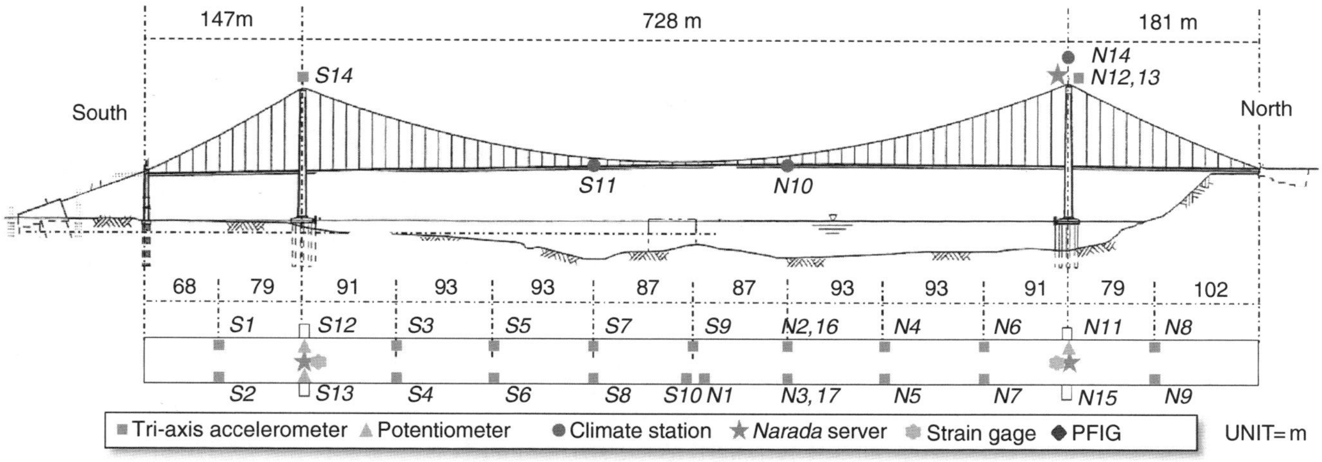

Fig. 8.1 Wireless sensor network deployment on the New Carquinez Suspension Bridge

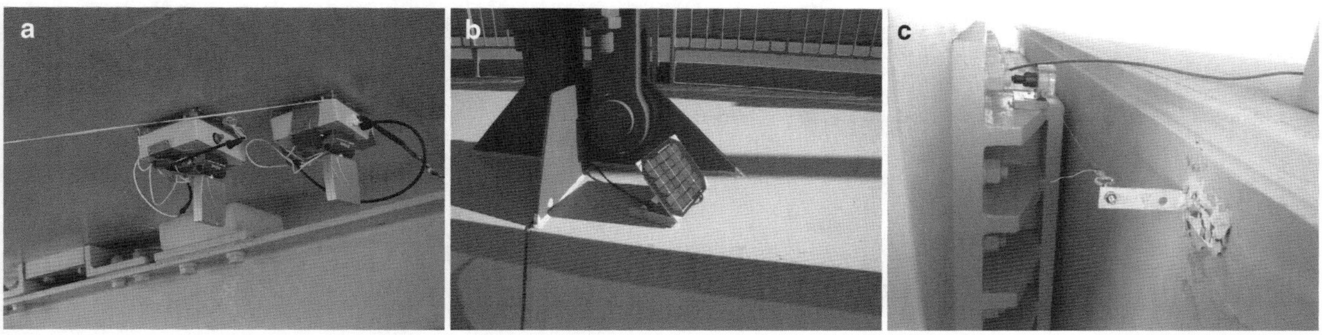

Fig. 8.2 Installation details of the New Carquinez Suspension Bridge long-term monitoring system: (**a**) typical *Narada* wireless sensor node beneath the deck girder in a magnetically mounted weather-proof enclosure; (**b**) small solar panel installed on the bridge top deck to power a *Narada* node; (**c**) string potentiometer at the deck-tower interface to measure longitudinal displacement

inexpensive single-board computer. The functionality of the server is to: (1) time synchronize the server clock using the network time protocol (NTP); (2) time synchronize the wireless sensors in each sub-network using a beacon time synchronization protocols previously developed for *Narada*; (3) command the wireless sensors to collect data; (4) locally store data on the on-board hard drive; and, (5) communicate data to the Internet via a cellular modem. The servers can collect data on either a schedule or upon user-demand; their default data collection occurs every 4 h with 8 min of bridge response data collected.

A massive database server termed *SenStore* has been created off-site to store the data and to expose application programming interfaces (APIs) to data clients seeking to use bridge data in their data processing algorithms. *SenStore* is a data management system designed for structural health monitoring (SHM) applications [4]. It uses a relational database to store bridge meta-data including structural component definitions, geometric details, sensor network data (e.g. sensor type and installation location), and structural analysis information (e.g., definitions needed for finite element modeling). Because the massive amounts of raw time history data from sensors are ill-suited for storage in a traditional relational database, a secondary database system based on the HDF5 file system is implemented in *SenStore* for the storage of sensor data. The relational database links directly to the HDF5 file system so that queries for sensor data can be handled by the *SenStore* server through the relational database. The server–client model implemented in the design of *SenStore* allows secure access to bridge data through a well-defined set of APIs. Computational tools are designed to perform autonomous system operation and to undertake data analysis. Tools for tracking sensor operation (i.e., fault detection), data visualization, and modal parameter extraction, among others, have been already created [7]. In this paper, the tools designed for automated modal extraction are primarily used to process bridge data.

8.3 Stochastic Subsapce Indentification (SSI)

In the system identification of civil engineering structures, output-only methods are a popular choice as there are many technical and practical issues that make use of direct input–output methods very difficult to employ. Specifically for large-scale bridges, it is difficult to obtain the input to the system such as live loads (e.g. heavy trucks). Other loads, such as wind loads, are difficult to fully quantify using discrete measurement of wind parameters. To circumvent the need to have load (input) information, several output-only system identification techniques have been adopted for the modal analysis of civil structures. Stochastic subspace identification (SSI) [8, 9] and frequency domain decomposition (FDD) [10] are common choices when performing the modal analysis of bridges to extract the modal properties [2, 4, 11]; other output-only methods also exist and have been successfully employed.

The vibration response of a discrete system with n masses (i.e., degrees of freedom) connected through springs and dampers can be described by the following second-order ordinary differential equation:

$$\boldsymbol{M}\ddot{U}(t) + \boldsymbol{C}_2\dot{U}(t) + \boldsymbol{K}U(t) = F(t) \tag{8.1}$$

where $\boldsymbol{M}, \boldsymbol{C}_2, \boldsymbol{K} \in \mathbb{R}^{n \times n}$ are the mass, damping and stiffness matrices, respectively, $F(t) \in \mathbb{R}^{n \times 1}$ is the excitation force vector, and $U(t) \in \mathbb{R}^{n \times 1}$ is the displacement vector of the system in continuous time t. For civil engineering structures, especially bridges, it is difficult to acquire and quantify the system input (load). As such, the ambient excitation of the system is typically assumed to have zero mean and broadband (i.e., white). For the reasons stated above, stochastic state-space model is introduced to describe the behavior of the dynamic system. For clarity, the SSI derivation provided by Peeters and Roeck [8] and Overschee and Moor [9] will be restated here; interested readers should consult these seminal works.

The state-space model of the continuous time system is described as follows:

$$\dot{x}(t) = \boldsymbol{A}_c x(t) + \boldsymbol{B}_c u(t) \tag{8.2}$$

$$y(t) = \boldsymbol{C}x(t) + \boldsymbol{D}u(t) \tag{8.3}$$

$$x(t) = \begin{bmatrix} U(t) \\ \dot{U}(t) \end{bmatrix} \tag{8.4}$$

$$\boldsymbol{A}_c = \begin{bmatrix} 0 & I_n \\ -M^{-1}K & -M^{-1}C \end{bmatrix} \tag{8.5}$$

$$\boldsymbol{B}_c = \begin{bmatrix} 0 \\ M^{-1}B_2 \end{bmatrix} \tag{8.6}$$

where $x(t) \in \mathbb{R}^{2n \times 1}$ is the state vector, $A_c \in \mathbb{R}^{2n \times 2n}$ is the system matrix and $B_c \in \mathbb{R}^{2n \times m}$ is the input matrix. Also, the output can be stated as:

$$y(t) = C_a\ddot{U}(t) + C_v\dot{U}(t) + C_d U(t) \tag{8.7}$$

$$\boldsymbol{C} = \begin{bmatrix} C_d - C_a M^{-1}K & C_v - C_a M^{-1}C_2 \end{bmatrix} \tag{8.8}$$

$$D = C_a M^{-1} B_2 \tag{8.9}$$

where $y(t) \in \mathbb{R}^{l \times 1}$ is the output vector and $\ddot{U}(t), \dot{U}(t)$ and $U(t) \in \mathbb{R}^{n \times 1}$ are the acceleration, velocity and displacement vectors, respectively, measured by transducers at l different locations. C_a, C_v and $C_d \in \mathbb{R}^{l \times n}$ are the output matrices for the state acceleration, velocity and displacement, respectively. Therefore $C \in \mathbb{R}^{l \times n}$ is the state-space model output matrix and $D \in \mathbb{R}^{l \times m}$ is the model direct transmission matrix.

Data is acquired at discrete time intervals $k\Delta t$, $k \in \mathbb{N}$, so the state-space model is reformulated in discrete-time as:

$$x_{k+1} = Ax_k + Bu_k + w_k \tag{8.10}$$

$$y_k = Cx_k + Du_k + v_k \tag{8.11}$$

where $x_k = x(k\Delta t)$ and Δt is the time-step. In the discrete-time state-space model, A is the discrete-time system matrix, and B is the discrete-time input matrix. u_k is the input of the system, which could be assumed to be zero in this application. As the system is stochastic, w_k and v_k are Gaussian white noise processes. Therefore the system is restated as:

$$\begin{bmatrix} x_{k+1} \\ y_k \end{bmatrix} = \begin{bmatrix} A & w_k \\ C & v_k \end{bmatrix} \begin{bmatrix} x_k \\ 1 \end{bmatrix} \tag{8.12}$$

The Numerical Algorithms for Subspace State Space System Identification (N4SID) [9, 12] toolbox is applied to determine the optimal system matrices $\boldsymbol{A}$ and $\boldsymbol{C}$ in this study. Block Hankel matrices, oblique projections and singular value decomposition are all important computational component to the N4SID algorithm for SSI. The output block Hankel matrix is constructed and defined as follows:

$$Y_{0|2i-1} \underline{\underline{\text{def}}} \left(\begin{array}{cccc} y_0 & y_1 & \cdots & y_{j-1} \\ \cdots & \cdots & \cdots & \cdots \\ y_{i-2} & y_{i-1} & \cdots & y_{i+j-3} \\ y_{i-1} & y_i & \cdots & y_{i+j-2} \\ \hline y_i & y_{i+1} & \cdots & y_{j-1} \\ y_{i+1} & y_{i+2} & \cdots & y_{i+j} \\ \cdots & \cdots & \cdots & \cdots \\ y_{2i-1} & y_{2i} & \cdots & y_{2i+j-2} \end{array} \right) \underline{\underline{\text{def}}} \left(\frac{Y_{0|i-1}}{Y_{i|2i-1}} \right) \underline{\underline{\text{def}}} \left(\frac{Y_p}{Y_f} \right) \tag{8.13}$$

while an alternative form is:

$$Y_{0|2i-1} \underline{\underline{\text{def}}} \left(\begin{array}{cccc} y_0 & y_1 & \cdots & y_{j-1} \\ \cdots & \cdots & \cdots & \cdots \\ y_{i-2} & y_{i-1} & \cdots & y_{i+j-3} \\ y_{i-1} & y_i & \cdots & y_{i+j-2} \\ y_i & y_{i+1} & \cdots & y_{j-1} \\ \hline y_{i+1} & y_{i+2} & \cdots & y_{i+j} \\ \cdots & \cdots & \cdots & \cdots \\ y_{2i-1} & y_{2i} & \cdots & y_{2i+j-2} \end{array} \right) \underline{\underline{\text{def}}} \left(\frac{Y_{0|i}}{Y_{i+1|2i-1}} \right) \underline{\underline{\text{def}}} \left(\frac{Y_p^{\,+}}{Y_f^{\,-}} \right) \tag{8.14}$$

Denotation of p means "past" and f means "future" variables. The joint space of the past and future can be defined as $W_p = [U_p{}^T \; Y_p{}^T]^T$. The oblique projection, $\boldsymbol{P}_i$, of Y_f along the row space of input U_f onto the row space of W_p is $P_i = Y_f /_{U_f} W_p$. As the input U_f is zero in the stochastic system, $\boldsymbol{P}_i$, is stated as:

$$P_i = Y_f / \, Y_p \tag{8.15}$$

The oblique projection is also equal to the product of the extended observability matrix, Γ_i, and the state sequence estimate, $\widehat{X}_i$, :

$$P_i = \Gamma_i \, \widehat{X}_i = \left[C^T \; (CA)^T \ldots \; (CA^{i-1})^T\right]^T \left[\widehat{x}_i \; \widehat{x}_{i+1} \ldots \widehat{x}_{i+j-1}\right] \tag{8.16}$$

The singular value decomposition of P_i is:

$$P_i = USV^T = (U_1 \; U_2) \begin{pmatrix} S_1 & 0 \\ 0 & 0 \end{pmatrix} \begin{pmatrix} V_1{}^T \\ V_2{}^T \end{pmatrix} = U_1 \; S_1 V_1{}^T \tag{8.17}$$

where S is the diagonalized singular values matrix, U is the left-singular vector matrix, and V is the right-singular vector matrix. Therefore, the extended observability matrix Γ_i and the state sequence $\widehat{X}_i$ are equal to:

$$\Gamma_i = U_1 \, {S_1}^{1/2}, \tag{8.18}$$

$$\widehat{X}_i = {S_1}^{1/2}{V_1}^T \tag{8.19}$$

Another similar oblique projection P_{i-1} is defined as:

$$P_{i-1} = \frac{{Y_f}^-}{{Y_p}^+} = \Gamma_{i-1} \, \widehat{X}_{i+1} \tag{8.20}$$

where Γ_{i-1} is defined as the matrix Γ_i without the last low. Therefore $\widehat{X}_{\mathrm{i}+1}$ can be calculated as:

$$\widehat{X}_{\mathrm{i}+1} = {\Gamma_{\mathrm{i}-1}}^{\dagger} P_{i-1} \tag{8.21}$$

The system equation is finally stated as:

$$\begin{pmatrix} \widehat{X}_{i+1} \\ Y_{i|i} \end{pmatrix} = \begin{pmatrix} A \\ C \end{pmatrix} \begin{pmatrix} \widehat{X}_i \end{pmatrix} + \begin{pmatrix} \rho_w \\ \rho_v \end{pmatrix} \tag{8.22}$$

where ρ_w and ρ_v are the noise residues. Since $\widehat{X}_{i+1}$ and $\widehat{X}_i$ are extracted from the SVD of P_{i-1} (8.20) and P_i (8.17) respectively, a least square method is used to find $\boldsymbol{A}$ and $\boldsymbol{C}$:

$$\begin{pmatrix} A \\ C \end{pmatrix} = \begin{pmatrix} \widehat{X}_{i+1} \\ Y_{i|i} \end{pmatrix} \widehat{X}_i^{\dagger} \tag{8.23}$$

The damping ratio and natural frequency of the *ith* mode can be calculated by converting the discrete-time eigenvalues of $\boldsymbol{A}$ to the continuous-time domain:

$$A = \Psi \Lambda \Psi^{-1} \tag{8.24}$$

where Λ is the diagonal matrix containing the eigenvalues of $\boldsymbol{A}$ and Ψ contains eigenvectors in each column. The mode shape vectors can be calculated as:

$$\Phi = [\Phi_1 \; \Phi_2 \cdot s \; \Phi_n] = C\Psi \tag{8.25}$$

As the measurements have noise and signal aliasing, not all the eigenvalues obtained are the corresponding eigenvalues of the modal frequencies of the structure. In addition, for low energy modes (e.g., torsional modes), the corresponding eigenvalues of these modes may not be able to be obtained from SSI. To autonomously extract the correct modal frequencies directly from the data collected from the sensor network, examination of the corresponding damping ratios of each mode are used to eliminate pure mathematical results. As for regular civil engineering structures, the damping ratio is typically lower than 5 %. If damping results from SSI is more than 10 %, it is highly likely that the corresponding mode has not resulted from the structural response and therefore such results should be discarded as a mathematical artifact.

8.4 Modal Analysis Results of New Carquinez Bridge

Modal properties of an invariant system are fixed in theory. However, for operational bridges, the operational environment changes thereby introducing some variance into the system; therefore the observed modal properties are not fixed. The New Carquinez Suspension Bridge, as one of the largest bridges in the Bay Area, experiences continual heavy traffic loads and large temperature variations. With the implementation of the long-term structural monitoring system, automated modal

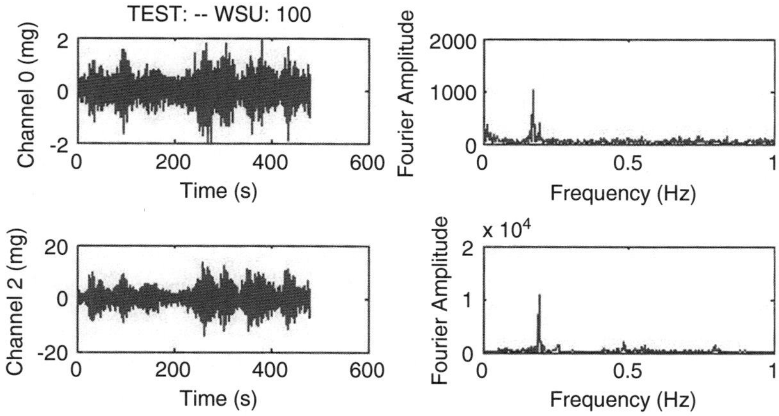

Fig. 8.3 Sample acceleration time-history data (*left*) and corresponding frequency domain representation (*right*)

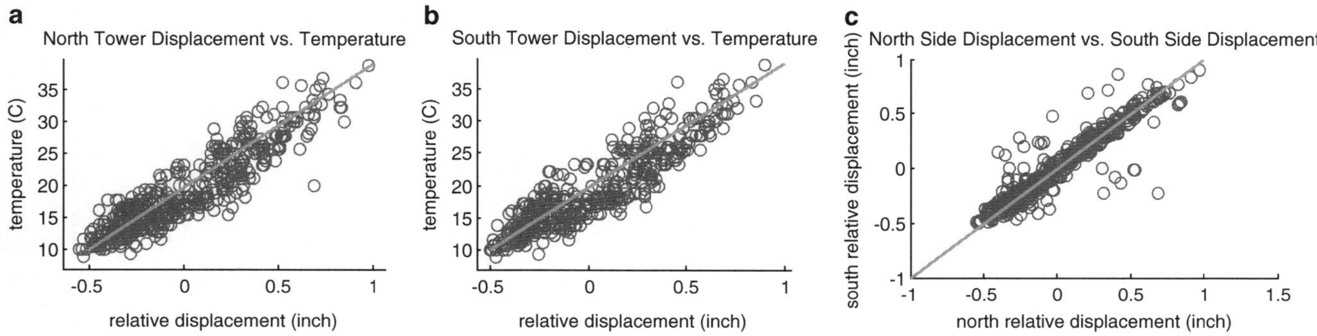

Fig. 8.4 Relationship between the bridge girder longitudinal displacements and temperature

analysis tools are used to extract modal properties at six different time periods over a day. Figure 8.3 is a typical plot of vertical acceleration data of the deck in the time and frequency domains; this type of data is used to extract modal properties using SSI. The relationship between modal frequencies and environmental variables is studied. In detail, the relationship between modal frequency and temperature, modal frequency and time of the day at different days in the week, and the relationship between horizontal movement of the bridge deck and temperature are all studied.

8.4.1 Relationship Between Modal Frequencies, Temperature and Traffic

Temperature change in the Bay Area varies in the range of 8–40 °C during a typical day in the summer. This large temperature change will impact the material properties of the steel girder box of the bridge. The orthotropic steel box girder of the New Carquinez Suspension Bridge is connected to the concrete footings at the wind tongues located at the towers. It is allowed to move in, longitudinal and vertical directions to compensate for imposed loads. To evaluate and monitor the performance of the bridge, the behavior of the deck movement is studied for the testbed structure. Data is collected from the string potentiometers installed longitudinally at the wind tongues (Fig. 8.2c) of both towers. As the initial position between the girder and the wind tongues is impossible to know, only the relative deck movement is obtainable as temperature varies (Fig. 8.4). This data verifies deck contraction with reducing temperature. The relative displacements from both sides of the bridge (at north and south towers) are also linearly related with a slope of 1 (Fig. 8.4c). As the way the string potentiometers are installed, the positive direction is defined as north for data from the north side and south for data from the south side. These facts indicate the bridge girder expands in the longitudinal direction when temperature increases and shrinks when the temperature decreases, which obey thermal expansion rules as expected. The contraction behavior due to temperature changes is symmetric when looking at the north and south sides of the bridge.

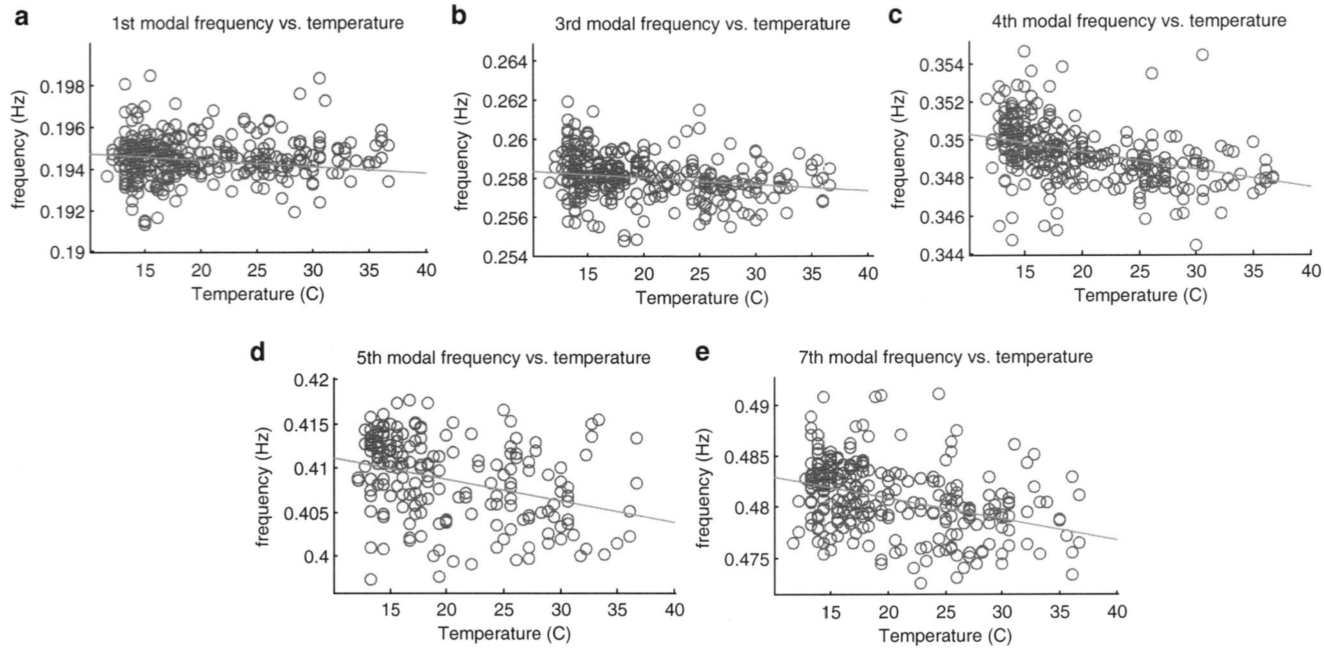

Fig. 8.5 Relationship of modal frequency and temperature: (**a**) 1st mode, (**b**) 3rd mode, (**c**) 4th mode, (**d**) 5th mode, and (**e**) 7th mode

To evaluate the performance of the bridge with the change of temperature, studies on the modal frequencies and temperature are performed. As a large scale structure, the New Carquinez Suspension Bridge has very low natural frequencies as shown in Fig. 8.3 with the first mode near 0.19 Hz. Due to the design of the structure, the second modal frequency is very close to the first mode, roughly 0.01 Hz a part. As a result, the second mode is not always autonomously detected by the automated modal analysis tools. In addition, the sixth mode is a torsional mode which has less energy and thus more difficult to detect. Therefore, the first, third, fourth, fifth and seventh modal frequencies are used for most of the analyses presented herein. Figure 8.5 shows the distribution of modal frequencies as a function of temperature. Modal frequency data presented in Fig. 8.5 are obtained using the SSI method and least square linear fitting is employed to present the trend of the data. It is shown that the 4th and 7th modes have stronger trending behavior of decreasing modal frequency as temperature increases as compared to the other three modes. However, decreasing modal frequencies as a function of temperature are generally observed for all of the modes. The modal frequency floating range is within 2 % of the mean value. As the observed data is not evenly distributed over the temperature range, it is hard to make a direct conclusion of the variance of the data for different temperature ranges.

Figure 8.6 shows the distribution of modal frequencies as a function of the time of the day. The sensor networks are configured to acquire data 6 times a day separated by 4 h apart. Data from roughly 60 consecutive days are included in this plot and spline of the mean values at each sensing point is fitted to show the data trend. From Fig. 8.6b, c, d, e, the data shows the trend that the modal frequencies decrease during the early half of the day and increase in the late half of the day. Besides the temperature variation (Fig. 8.6f), change of the traffic load is another important event that happens during the day which would also temporarily change the performance of the bridge because traffic imposes dynamic loads and increases the weight of the structure. It is difficult to acquire accurate traffic load data, but a good estimation could be made based on the relationship of traffic and time of the day. Traffic loads are higher during the day time compared to the night time and the peaks occur at rush hour (8 a.m. and 5 p.m.). In Fig. 8.6d, e, the frequencies seem to dip lower at rush hour as compared to frequencies from other non-rush hour periods. In Fig. 8.7 the mean of the normalized frequency data of each mode and the mean of normalized temperature data at each sampling time are presented on the same figure. The amplitudes of the modal frequency data curves are amplified by 20 times to exaggerate the trend. Modal frequency data are fitted by splines. Only the first mode data behaves differently than the other modes in the trending over time of the day. The rest of the modes (third, fourth, fifth and seventh) have similar general trending with frequency declining while the temperature is rising and vice versa. Therefore, the influence of temperature and traffic loads on the New Carquinez Bridge is certain and not negligible. For a certain mode, the bridge tends to operate in a frequency lower than the natural frequency as the temperature and/or the traffic loads increase and tends to operate in a frequency higher than the natural frequency as the temperature and/or the traffic loads decrease.

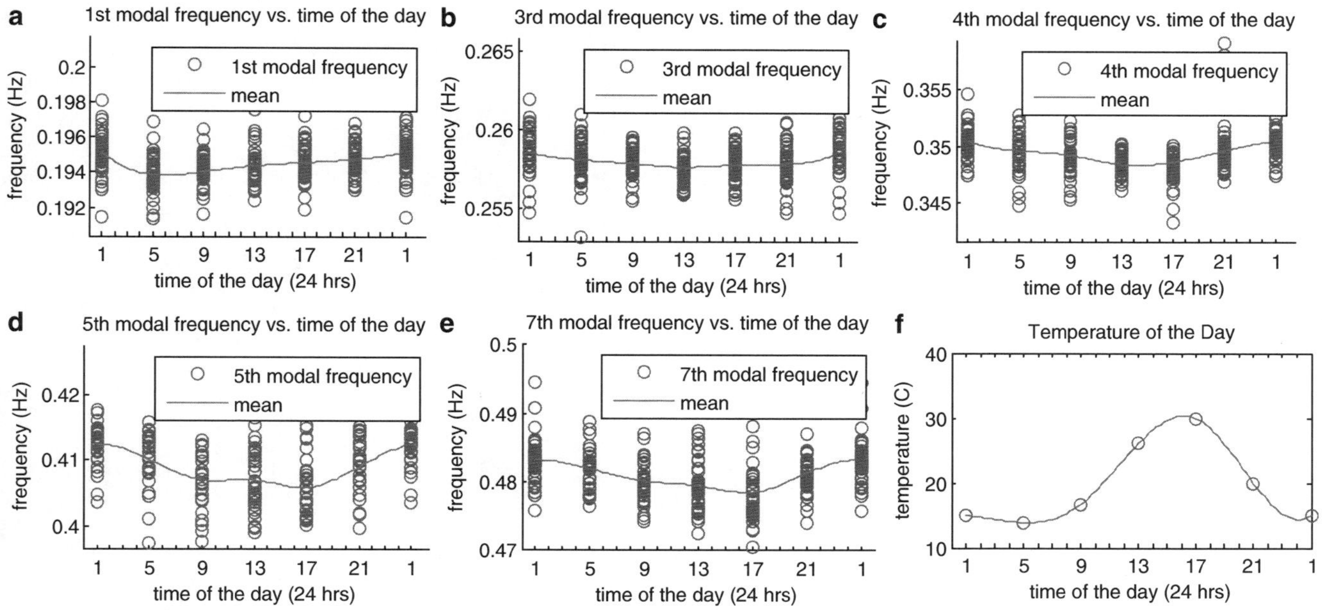

Fig. 8.6 Relationship of modal frequency and time of the day: (**a**) 1st mode, (**b**) 3rd mode, (**c**) 4th mode, (**d**) 5th mode, (**e**) 7th mode, and (**f**) temperature

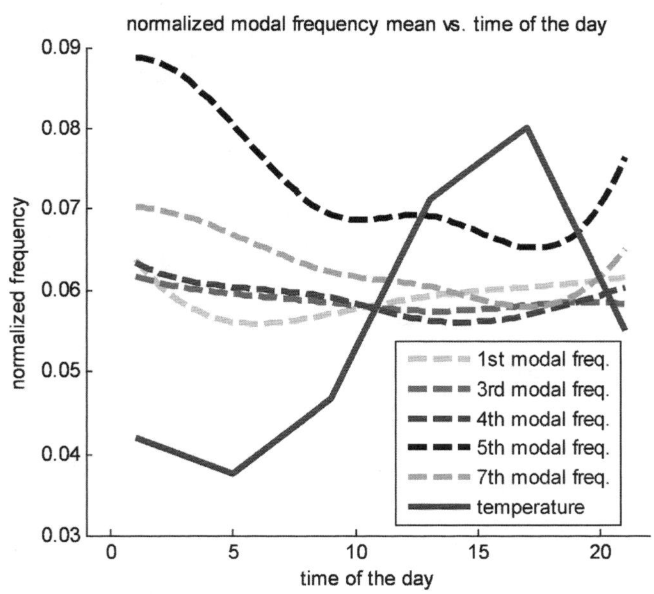

Fig. 8.7 Mean of the modal frequency distribution as a function of time of day

8.5 Conclusions

This paper discussed the modal analysis methods and results of a long-term structural monitoring study on the New Carquinez Suspension Bridge. With an automated modal characteristics extraction tool, modal frequencies are successfully obtained using the output-only SSI method. Second and sixth modes have low levels of energy and are therefore difficult to identify. Therefore the first, third, fourth, fifth and seventh modes are used to study the relationship between modal frequencies and environmental variables. For the third, fourth, fifth and seventh modes, the observed modal frequencies tend to decrease as the temperature increases. The fourth and seventh modes have stronger decreasing linear relationships with the temperature. Moreover, the traffic also has an impact on the performance of the bridge; the modal frequencies tend to decrease as the traffic loads increase. In addition, the longitudinal movement of the orthotropic steel box girder is found to behave linearly with temperature. Future work is exploring modal information collected more frequently during the day (e.g., every hour) to verify the identified reduction in modal frequencies due to traffic loads.

Acknowledgments The authors would like to gratefully acknowledge the generous support offered by the U.S. Department of Commerce, National Institute of Standards and Technology (NIST) Technology Innovation Program (TIP) under Cooperative Agreement Number 70NANB9H9008. Additional support was provided by the University of Michigan and the California Department of Transportation(Caltrans). Support and guidance in the use of SenStore provided by Gwen van der Linden (SC Solutions) and Professor Atul Prakash (University of Michigan) are greatly appreciated. On site assistance with sensor installation on the New Carquinez Bridge by Caltran personnel, especially Mr. Edward Thometz, are also greatly appreciated.

References

1. Shakal A, Huang M (1989) Overview of the strong motion instrumentation program. In: Seminar on seismological and engineering implications of recent strong-motion data. Sacramento, California
2. Cross E, Koo K, Brownjohn J, Worden K (2012) Long-term monitoring and data analysis of the Tamar bridge. Mech Syst Signal Process. February 2013, 35(1–2):16–34
3. Ni Y, Hua X, Fan K, Ko J (2005) Correlating modal properties with temperature using long-term monitoring data and support vector machine technique. Eng Struct 27(12):1762–1773
4. Kurata M, Kim J, Lynch J, Linden G, Sedarat H, Thometz E, Hpley P, Sheng L (2012) Internet-Enabled Wireless Structural Monitoring Systems: Development and Permanent Deployment at the New Carquinez Suspension Bridge. J. Struct. Eng. http://ascelibrary.org/doi/abs/10.1061/%28ASCE%29ST.1943-541X.0000609
5. Kurata M, Kim J, Zhang Y, Lynch J, van der Linden G, Jacob V, Thometz E, Hipley P, Sheng L (2011) Long-term assessment of an autonomous wireless structural health monitoring system at the new carquinez suspension bridge. In: SPIE, San Diego
6. Swartz R, Jung D, Lynch J, Wang Y, Shi D, Flynn M (2005) Design of a wireless sensor for scalable distributed in-network computation in a structural health monitoring. In: 5th international workshop on structural health monitoring, Stanford
7. Zhang Y, Kurata M, Lynch JP, van der Linden G, Sadarat H, Prakash A (2012) Distributed cyberinfrastructure tools for automated data processing. In: SPIE, San Diego
8. Peeters B, de Roeck G (1999) Reference-based stochastic subspace identification for output-only modal analysis. Mech Syst Signal Process 13(6):855–878
9. Van Overschee P, De Moor B (1996) Subspace identification for linear system: theory, implementation, applications. Kluwer Academic, Dordrecht
10. Brincker R, Zhang L, Andersen P (2001) Modal identification of output-only systems using frequency domain decomposition. Smart Mate Struct 10(3):441–445
11. Ni Y, Fan K, Zheng G, Chan T, Ko J (2003) Automatic modal identification of cable-supported bridges. In Smart materials and structures 10, no. 3 (2001):441.
12. Kim J, Lynch J, (2011) Comparison study of output-only subspace and frequency-domain methods. In: Civil Engineering Topics, Volume 4, pp. 305–312.

Chapter 9
Static and Dynamic Monitoring of Bridges by Means of Vision-Based Measuring System

Giorgio Busca, Alfredo Cigada, Paolo Mazzoleni, Marco Tarabini, and Emanuele Zappa

Abstract Civil structure static and dynamic monitoring is a key activity for both safety and maintenance purposes. In this paper the use of cameras to monitor a bridge response to train transits is being considered and results are presented and compared with a reference measurement, provided by a laser interferometer. A camera is a non contact measurement device, having mainly two advantages: first of all the effort required to prepare the set-up is very low, because the camera is fixed in the proximity of the structure and the targets to be mounted on the bridge (if required) do not need any cable connection for both power supply and signals. Secondly cameras can measure the displacements of multiple targets in the field of view. The leading idea of the paper is to use image acquisition and processing (Pattern Matching and Edge Detection) not only to measure the displacement of a few targets but also to grab images from a wide structure portion in order to recover displacements of a large number of points in the field of view. The extreme final solution would be having wide area measurements with no targets, to make measurements really easy, with evident advantages, but also with some drawbacks to be fully comprehended.

Keywords Vision • Bridges • Dynamics • Monitoring • Uncertainty

9.1 Introduction

Static and dynamic testing of structures is at the same time a fundamental and critical operation, mainly due to the fact that this type of testing is required for both safety and maintenance purposes. In the case of bridges, the need to periodically assess the structure stability and runability makes this need relevant. Together with the regional railway company of Lombardy (Ferrovie Nord Milano), it has been decided to consider a 50 m long steel trussed bridge, as a laboratory bridge, to be used for testing vision-based motion monitoring techniques to get the best bridge response estimation and to gain awareness of the limits related to this approach. The considered structure is an old steel trussed bridge crossing a river; it allows good access from both sides, therefore responding to the requirements of being a sort of 'laboratory' bridge in which any change in the measurement parameters are possible to obtain without too difficult operation. Trains travel on the bridge at a rather low speed, and also dynamic phenomena involving the whole bridge are confined to the low frequency range. The main goal of this work is to define the performances of vision-based displacement measurement techniques in terms of sensitivity, resolution, uncertainty, in the harsh field environment, which might strongly deviate from the behaviour defined during laboratory testing.

In this work vision approaches have been applied together with more traditional measurements: among these a single-point laser interferometer has been adopted as the reference, at mid span, to measure the effects of train pass by in terms of structural static and dynamic displacements (also the bandwidth definition is among the aims. The known advantages of camera-based measurement devices, that makes them attractive for the application of bridge monitoring, include: remote

G. Busca (✉) • A. Cigada • P. Mazzoleni • M. Tarabini • E. Zappa
Dipartimento di Meccanica, Politecnico di Milano, via La Masa 1, Milan, Italy
e-mail: giorgio.busca@mail.polimi.it; alfredo.cigada@polimi.it; paolo2.mazzoleni@mail.polimi.it; marco.tarabini@mail.polimi.it; emanuele.zappa@polimi.it

A. Cunha (ed.), *Topics in Dynamics of Bridges, Volume 3: Proceedings of the 31st IMAC, A Conference on Structural Dynamics, 2013*, Conference Proceedings of the Society for Experimental Mechanics Series 38, DOI 10.1007/978-1-4614-6519-5_9, © The Society for Experimental Mechanics, Inc. 2013

contactless monitoring (i.e. no critical problems in setting up the measurement system, especially in the case of bridges with frequent train passages), possibility to perform a multipoint measurement with a single camera, quick and simple measurement device set-up (no transducers nor cables on the structure), no need of access in critical areas, especially below the bridge to fix transducers. The mentioned aspect are even more attractive in the case of bridges crossing rivers or deep valleys.

An additional advantage in using a camera for displacement measurements is that every row or column in the pixel matrix can be considered a sensor on its own: so the complete camera corresponds to a very high number of parallel sensors, allowing for distributed sensing or giving the chance to exploit this redundancy to improve measurements reliability. Due to this reason, cameras are usually referred to as 'dense' sensors. However, cameras show the common problems of relative displacement transducers and have a restricted frequency bandwidth.

9.2 State of the Art

Thanks to the development of digital cameras, to the growth of computer processing capabilities and to the new image processing libraries, the use of vision-based measuring systems to measure the vibration of targets has become popular in the last years, with particular reference to the vibration monitoring of civil structures.

One of the issues in this kind of measurement is the type of target used for the measurement itself: in most the cases planar black and white targets are being attached to the structure in order to improve the measurement technique reliability and to reduce result uncertainty of the [1–5]. In other circumstances, the natural texture of the structure under investigation can be used for the measurement [6–8]. The main advantages of the markeless case are that the preparation of the set-up is much faster and that it is not necessary to have access to the structure to be monitored, no permissions, no target maintenance.

In [1] and [9] bridge vibration measurement is carried out using fit-to-the-purpose targets fixed to the structure; in this case the target is constituted by one black circle in white background, while in [2] a planar target with four circles is used. In [3] cross-shaped targets are used and the viewing system is equipped with an additional reference system, which decreases the sensitivity of the camera basement to ambient vibrations. Two types of targets are used in [4]: ring-shaped and random ones; in this case multiple target are measured contemporarily with a single camera, allowing multi-point measurements. In some applications active targets are used; for example in [5] LEDs are used for suspension bridge vibration monitoring. Markerless solutions are proposed in [6, 7] and [8], to monitor cable vibrations in cable-stayed bridges also in the case of power head transmission lines. In [10] vibration measurements obtained through image acquisition and processing are used to develop a modal analysis of a simple structure.

9.3 Design of the Vision-Based Measurement System and Testing Layout

In vision-based bridge monitoring a compromise must be looked for between two needs: on one side a wide field of view can theoretically allow to get the whole dynamic deformed shape, but this fights against the need to get a reasonable resolution in terms of pixel/mm. If the will to get the complete bridge deformed shape pulls towards the need of having a wide view, there is the serious risk that this choice impacts on resolution, so that the peak to peak vibration amplitude is confined within a few pixels, thus worsening the signal-to-noise ratio. Due to this reason, in the past, most dynamic measurements have been performed on small structure portions, trying to increase contrast, developing fit-to-the-purpose targets to be fixed to the structure, being rather close to the target and assuming that motion is rigid, to rely on the redundancy offered by the grabbed images to increase the reliability of results: in fact a single image allows to measure the motion of multiple targets, therefore in case of rigid motion the average of the displacements estimated by different targets allows to obtain a more robust motion estimation.

The steps faced in this paper are to which extent it is reasonably possible to move from 1D measurements to 2D dense measurements along a bridge span by using cameras. Once fixed the limits, this method can be really attractive, as it works with no sensors on the bridge, no power supply, in a simple approach manageable by whichever worker. This allows to get more dense checks (since the measurement setup can be easily assembled and disassembled), dramatically changing all the approaches to maintenance and testing.

The chosen test bed is a 50 m long steel trussed railway bridge crossing a river close to a village named Erba (Fig. 9.1a).

Trains run on bridge at low speed during the whole day, approximately every 30 min. In this work, vision based measures will be exploited in order to quantify the bridge sag during the trains pass-by: a movie of the structure during every train

Fig. 9.1 Tested structure (**a**), laser interferometer at mid span (**b**) and camera framing the bridge (**c**)

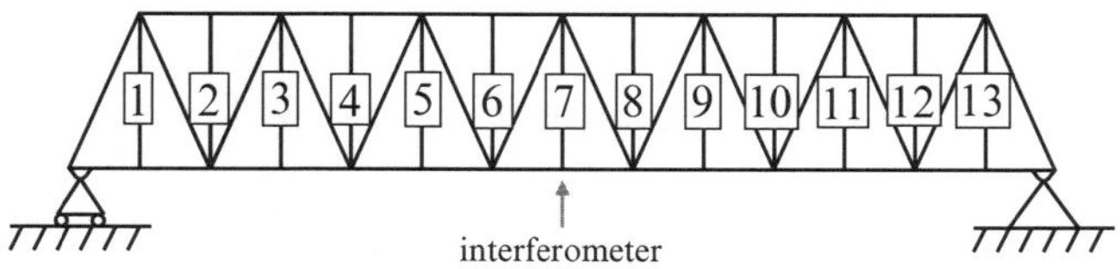

Fig. 9.2 Measuring layout

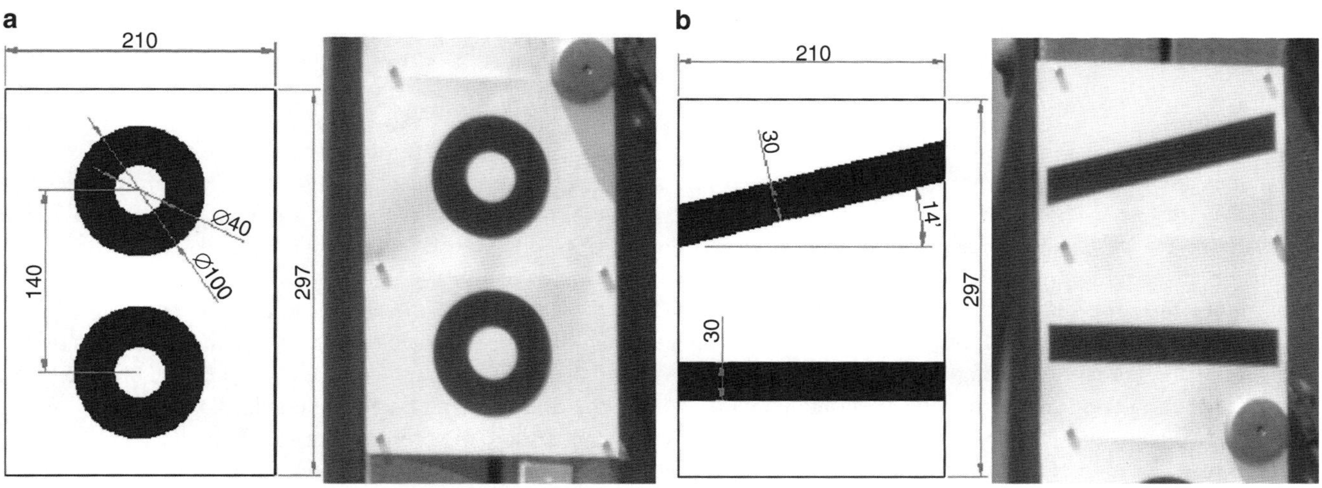

Fig. 9.3 Targets fixed on the bridge: (**a**) for pattern matching and (**b**) for edge detection analyses

crossing has been acquired and subsequently analyzed in order to estimate the vertical displacement at different points along the bridge main span. The movies are acquired by means of a standard user-consumer digital camera (Canon Legria HF 21, 1920 × 1080 px, 25 fps, Fig. 9.1c) to prove that no special hardware is required for these measurements.

The upper and the lower horizontal beams of the structure are connected by 13 vertical trusses on each side (Fig. 9.2, where a schematic sketch of the structure is reported). These trusses represent the points where the bridge vertical displacement will be measured. All the analyses will be carried out by two different image processing codes: a pattern matching and an edge detection algorithm. On every vertical truss, two targets were fixed, both printed on A4 sheets (210 × 297 mm).

The first target (Fig. 9.3a) is composed by a couple of round blobs and it will be used to both fix the scaling factor (i.e. px to mm conversion) and track the pattern matching sequence. Local scaling factor is computed for every measuring point in every movie, knowing the physical centre to centre distance of the two circles and estimating, by means of standard blob analysis, their distance (in px) in the first acquired frame of every movie. The second target (two black lines, Fig. 9.3b) will be analyzed by an edge detection algorithm. Only data extracted by the evaluation of the tilted line edge will be presented in this work: edge detection algorithms, in fact, have proven to work better when the edge is not aligned to the grid of the camera sensor [11, 12]. In order to provide a low uncertainty reference to these measurements, a laser interferometer (resolution = 2.56 μm) has been placed at the bridge mid span (Fig. 9.1b). The displacement measured by the interferometer can be assumed equal to that of target 7 in Fig. 9.2 and it will be used to compute the measurement uncertainty associated to the vision-based approach. By varying the portion of the bridge framed by the camera (i.e. the zoom level and the distance between the camera and the structure), it has been possible to study the relationship between the scaling factor and the obtained measurement uncertainty.

9.4 Experiments

In this section the results of the Erba bridge tests will be reported. The first series of tests aims at verifying the measurement performance in frontal view, i.e. with the camera placed at a side of the bridge, with the optical axis normal to the train direction and pointing several targets fixed onto the steel structure (Fig. 9.1c). The second series of tests has been carried out to verify the measurement performance in case a camera is being used to detect the bridge displacement without any target fixed to the structure, relying on the structure natural texture. These tests cannot be considered a real and complete innovation in case of controlled laboratory conditions: a smaller number of reported cases relates to field measurements. It must be noted that these tests are being performed under uncontrolled environmental conditions and are applied to a civil structure monitoring, where the image analysis must deal with the specific requirements of this practical application. In comparison with other similar works about real structures [8, 9], this one is focused on the uncertainty qualification of these techniques applied under uncontrolled environmental conditions.

9.4.1 Frontal Measurements with Target Fixed to the Structure

These measurements have the purpose to qualify the results obtained by the image analysis when the camera is placed in front of the bridge side. As explained in the previous section, several targets, one every 3.65 m, have been placed on the structure (Fig. 9.3). The tests specifications are summarized in Table 9.1: for every test the value of the px/mm ratio is given for the targets acquired by the camera. During test 1 the camera was near the structure and the zoom parameter has been set to have only one single target enclosed in the field of view. Under these conditions the highest image resolution in terms of px/mm ratio has been achieved, but for just one measurement point. All the further tests have been made by increasing the number of acquired targets so that a wider description of the vertical displacement of the bridge deck could be obtained by a single video. However, increasing the target number has the drawback of decreasing the image resolution in terms of the px/mm ratio, which means to worsen the measurement accuracy. For this reason, the presented results will consider tests with no more than five targets framed by the camera; further increase in the mm/px scaling leads to unacceptable measurements uncertainty.

The aim of these tests was to evaluate how the accuracy of the estimated displacements is affected by the target number and consequently by the image resolution. This evaluation has been carried out on a real big structure and with uncontrolled environment conditions, which is the situation in which the whole process is expected to operate for structural monitoring purposes. For every target the pattern matching technique is applied to the two blobs, whereas the edge detection is applied to the tilt line and a mean value of its position at every frame is extracted (Fig. 9.3). In this way it is possible to estimate the target displacement as a function of time.

Figure 9.4a shows the results obtained for test 1 (measurement point #7), where x axis is time and y axis gives the vertical displacement at mid span. The bridge vertical displacement, due to the train passage, which is between 7 and 8 mm can be

Table 9.1 Frontal tests: framed targets and local scaling factor (px/mm)

	Target				
Test # (number of framed targets)	#3	#4	#5	#6	#7
1					1.561
2				0.354	0.337
3			0.195	0.191	0.187
4		0.132	0.133	0.133	0.133
5	0.107	0.109	0.110	0.110	0.108

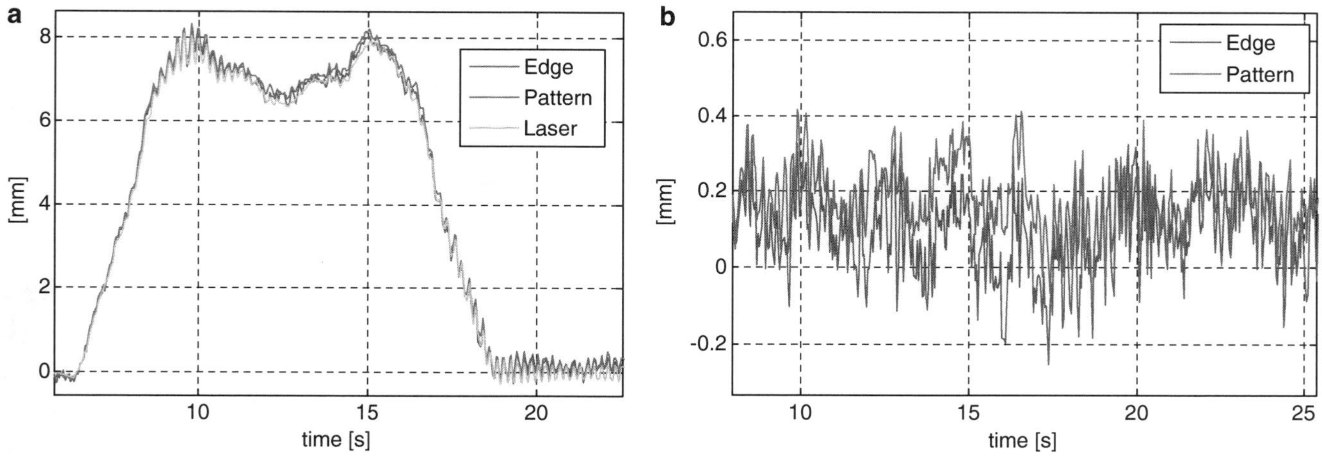

Fig. 9.4 Displacements measured at point 7 (**a**) and discrepancies between camera and laser (**b**)

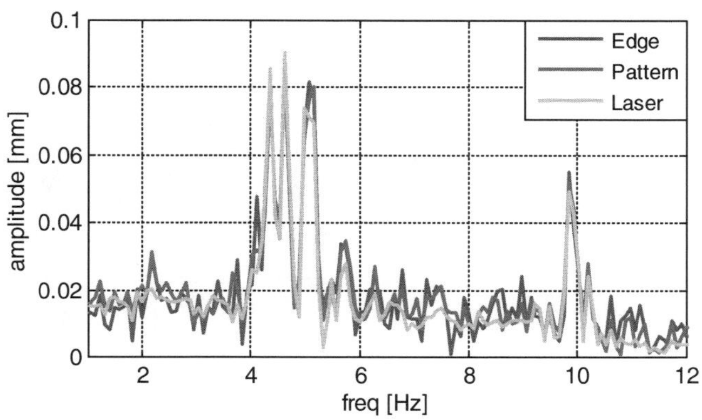

Fig. 9.5 Moduli of the spectra of the bridge sag computed at point 7: test 1

appreciated. The green line identifies the laser displacement measurement which can the considered the reference for the camera qualification. The red and blue lines represents the displacements estimated with the edge detection and pattern matching techniques respectively. The figure shows a good correspondence among data obtained with the three techniques. A direct comparison in the time domain between the displacement measured by the camera (the edge and the pattern analysis) and the reference (the laser) is given in Fig. 9.4b, where the maximum discrepancy is around the absolute value of 0.4 mm (this is the method uncertainty, as data are the same).

In Fig. 9.5 the moduli of the spectra (1–12 Hz) of the bridge sag computed starting from camera based measurements are compared with the one of the reference interferometer. Both edge detection and pattern matching are able to provide a reliable estimation of the dynamic displacement of the structure: the peak frequencies are correctly identified and their amplitudes are properly estimated. An overall slight overestimation of the structure vibrations can be notice in all the harmonic components characterized by low levels of vibration: this is a consequence of the lower signal to noise level of the camera based measurements with respect to the reference interferometer.

The same evaluations have been made for the other tests, but a direct comparison with the reference signal is possible only for target #7 (Fig. 9.2) because no reference transducers are available for the other tested bridge sections. However, it should be noted that the camera is mounted with the optical axis almost normal to the bridge, therefore, in each test, the scaling factor is nearly the same for all the targets; thanks to this consideration we can assume that the validation for the measurement at point #7 is valid also for the other targets. In order to give a complete description of what can be measured with a single video, all the estimated displacements of test 5 are shown together in Fig. 9.6. In this case data are obtained by the pattern matching algorithm applied to all the five targets. The results give a description of the bridge sag at different positions along the deck, and data are consistent with simple models of a beam supported at both ends, not being possible a direct check with a reference measurements for all points.

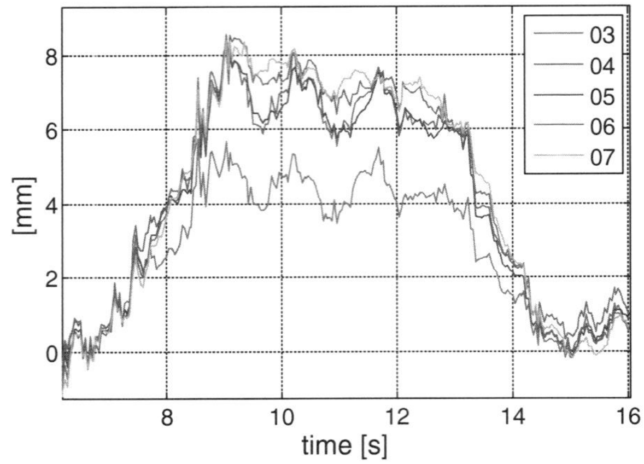

Fig. 9.6 Displacements measured from point 03 to point 07 in test 5 applying pattern matching algorithm

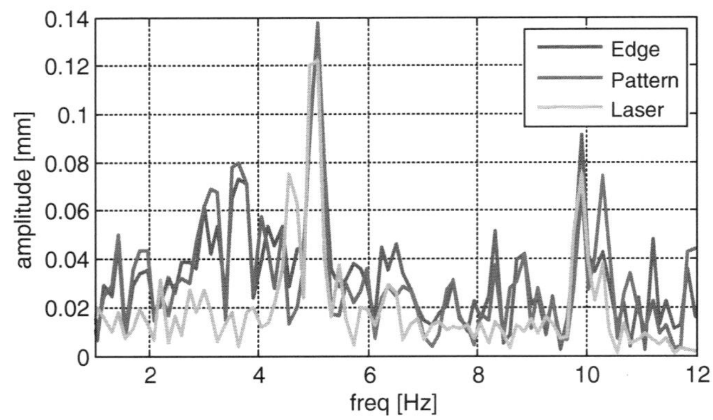

Fig. 9.7 Moduli of the spectra of the bridge sag computed at point 7: test 5

It can be qualitatively noticed a higher noise level in the single point measure, as a consequence of the lower px to mm scaling factor. The same conclusions can be derived by the comparison of the signal spectra estimated by the camera with respect to the one measured by the interferometer at point #7 (Fig. 9.7): as in the previous example, the camera based approaches are correctly able to identify the peak frequencies but a higher noise level can be notice as a consequence of an increased framed area.

Bringing attention to the measurement qualification, the Fig. 9.8 summarizes the results obtained from the tests, for the position of #7; in the figure the RMS value of the discrepancy between the reference signal and the displacement estimation is shown. The evaluation has been performed only for the time record corresponding to the train transit. The values are plotted as a function of the image resolution in terms of the mm/px ratio. The data quantify the trend of the measuring uncertainty with respect to the scaling factor. The results, as expected, show an increasing trend.

Figure 9.9 shows the RMS dynamic component (the standard deviation) and the RMS static component (the mean value) which are related to the RMS value by the formula:

$$RMS = \sqrt{\frac{1}{N}\sum \Delta^2} = \sqrt{\left(\mu(\Delta)^2 + \sigma(\Delta)^2\right)}.$$

where $\mu(\Delta)$ and $\sigma(\Delta)$ are respectively the mean and the standard deviation of the discrepancies between camera and reference signal in correspondence of a fixed target.

The analysis of the mean and standard deviation trends help to understand the RMS behaviour in Fig. 9.8. The standard deviation shows a clear linear increasing trend up to about 5 mm/px resolution value, whereas the mean takes random values between −0.20 and 0.20 mm. This means that the uncertainty linked to the dynamic component of the bridge vibrations is strongly affected by the image resolution, whereas the uncertainty of the static measure of displacement due to the train mass depends on the uncontrolled biased errors (mainly uncertainty in the mm/px ratio).

Finally, in Fig. 9.10 the mean and the standard deviations of the discrepancies between the interferometer and vision data are shown in the case of no train on the bridge. The goal of this analysis is to analyse the uncertainty in this condition and compare it with the measuring uncertainty obtained during the train transit. As it can be seen in Fig. 9.10, the values and the trend of the data match well with those of the previous figures. The conclusion is that the measurement uncertainty does not

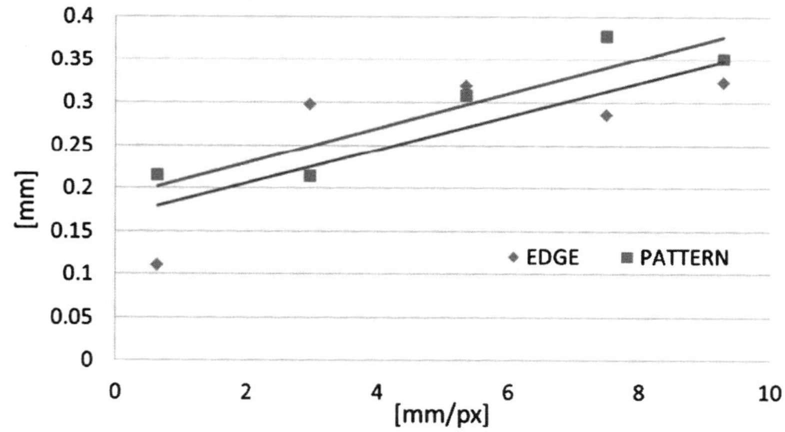

Fig. 9.8 RMS of the discrepancy of target p07 as a function of resolution

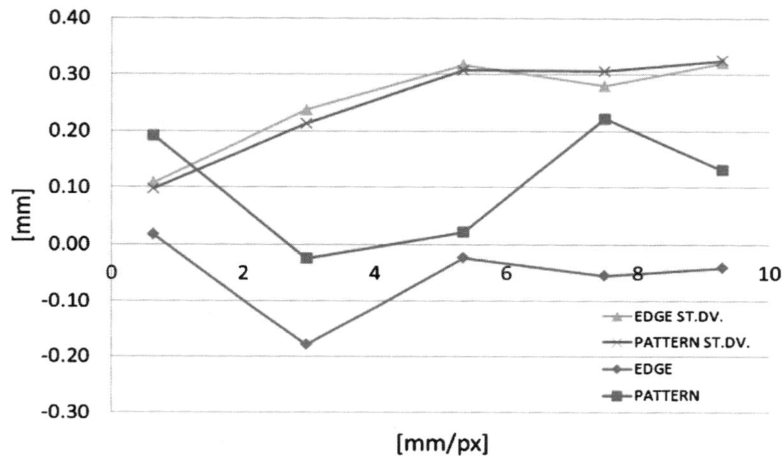

Fig. 9.9 Mean and standard deviation of the discrepancy of target p07 as a function of resolution

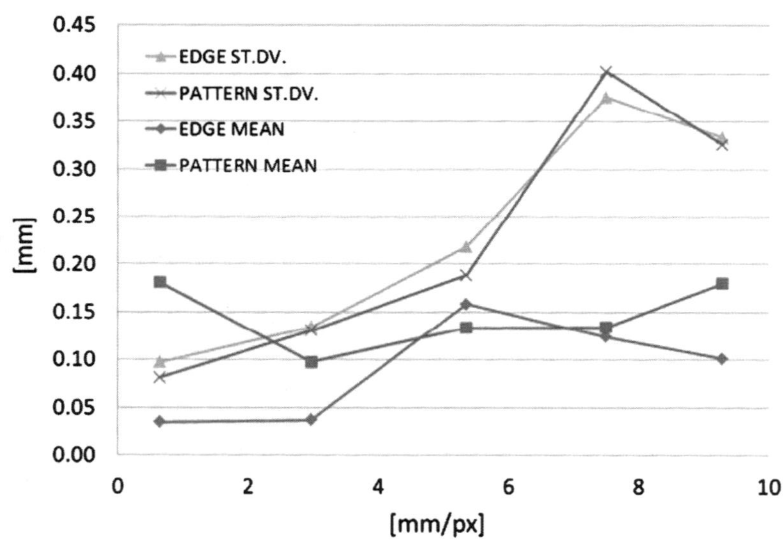

Fig. 9.10 Mean and standard deviation of the discrepancy in correspondence of no train transit

Fig. 9.11 Five patterns test, target-less analyses: first frame of the movie and tracked features

depend upon the vibration condition experienced by the bridge; on the contrary, it is only due to the image resolution and to errors in the calibration process; it is consequentially possible to estimate the accuracy of camera based measurements by simply framing the unloaded tested structure, without a real need for calibrated external reference.

9.4.2 Targetless Analysis

The analyses presented in the previous paragraph were based on the presence of fit-to-the-purpose targets at the measuring points. This approach presents certain advantages but it requires to access the structure before the test execution, which cannot always be guaranteed in real applications. The proposed techniques can work even without any target, relying on the natural texture of the structure: the analyses of the resulting performances in the targetless conditions will be presented in this paragraph.

Targets on the measuring points present two main advantages: on one hand, a framed known geometry makes the computation of the mm/px scaling factor straightforward; on the other hand, it is possible to design sharp edges and high contrasted patterns, basically increasing the signal to noise ratio of the measurement process.

In order to present an unbiased comparison with respect to the analyses performed using targets, only the second aspect (i.e. the possibility to obtain sharp and high-contrast target) will be investigated, maintaining the previously computed scaling factor.

The comparison has been performed only close to point #7, near one of the five test targets (Fig. 9.11), where the laser interferometer acquires the bridge sag; in this test the camera frames the bridge with a scaling factor of 0.108 mm/px.

Two different features have been extracted and tracked in the movie (Fig. 9.11). The first is a horizontal strip due to the presence of a white tape stuck to the bridge. This feature represents a less intrusive pattern with respect to that analyzed in the previous paragraph. This feature has been tracked with both the edge detection and the pattern matching algorithms. The extracted pattern dimension is 18×83 px and edge detection data have been averaged on the same width (18 px). The second feature is a riveted section of the bridge. Due to the lack of evident edges, only pattern matching can be applied for this analysis, on the 57×58 px pattern of Fig. 9.11.

In Fig. 9.12 the discrepancies with respect to the data recorded by the interferometer during the train transit are graphed and the previously presented synthetic data are reported. Considering the first pattern (tape), a slight increase in the standard deviation of the discrepancies, with respect to the data obtained from the target, can be noticed for both edge detection (from 0.31 to 0.36 mm) and pattern matching (from 0.33 to 0.35 mm) analyses. Considering the small differences in the measurement performances with respect to the previous presented data, this approach can still be considered attractive in cases where either the presence of the operator on the structure has to be minimized or the surface presents inherent textures comparable to the tested one.

The third curve describes, conversely, a more critical situation: the rivet plate connecting the two beams is characterized by very low gradient in its textures and this results, in the analysis, in a higher measuring uncertainty. The standard deviation of the discrepancy is about 0.82 mm (more than twice the one obtained with the tape), with peaks that reach 2.5 mm on an 8 mm displacement.

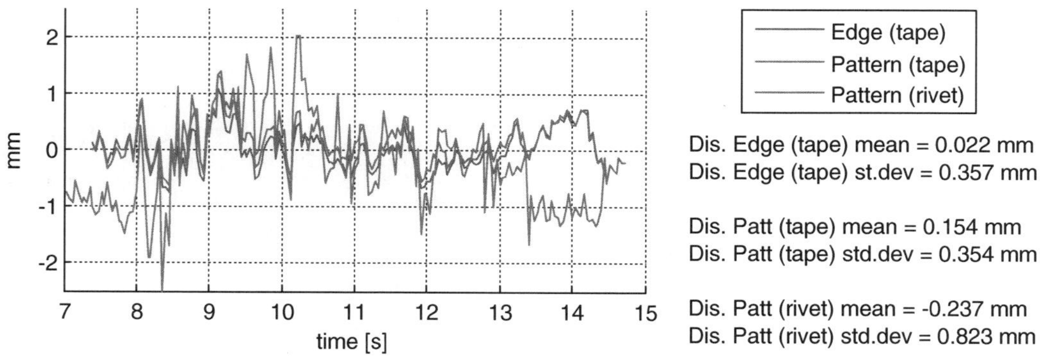

Fig. 9.12 Five patterns test, target-less analyses: discrepancies with respect to the interferometer data

9.5 Concluding Remarks

Static and dynamic monitoring of civil structures represents a crucial step in serviceability and health assessment. Vision-based measuring techniques are in these years gaining more and more importance, due to their advantages in terms of density of measured points and static and dynamic capabilities. In theory, nowadays image processing algorithms allow to appreciate displacement fields with an accuracy that can reach a few thousandths of pixel, but in real applications the performances strongly decreases due to external uncertainty contributions, such as lighting conditions, sensor noise, camera vibration and scaling factor estimation. As a consequence their application results in a compromise between field of view and accuracy. In this work, the relation between measurement uncertainty and spatial resolution is studied in a real environment. Two different state of the art image processing algorithms, edge detection and pattern matching, are exploited to estimate the vertical sag of a railway bridge subjected to train transit. The collected data are compared with a low uncertainty reference, a laser interferometer transducer, in order to quantify the vision-based measuring uncertainty.

At first, the displacement is measured in correspondence of fit-to-the-purpose targets mounted on the structure: the root mean square of the discrepancy between camera-based measurements and reference transducer shows an approximately linearly increasing trend with respect to the setup scaling factor. The RMS of the discrepancy has been decomposed into its two components: the standard deviation, representing the random component of the discrepancy and characterized by a deterministic trend if plotted against the scaling factor, and the mean discrepancy (bias), that, on the contrary, does not show any evident trend. Furthermore, it has been proven that the measuring uncertainty estimated in static condition (no loading of the structure) is able to correctly quantify the measuring system uncertainty, suggesting an easy way to estimate such a parameter in an on-the-field application.

In the last part of the paper the displacement of the bridge deck is measured without using the targets but relying only on the natural texture of the bridge. In this conditions the measurement reliability is strongly affected by the structure texture contrast, however, in favourable conditions, small differences in the measuring performances are found with respect to the measurement with target mounted on the bridge. The markerless approach can still be considered attractive in cases where either the presence of the operator on the structure has to be minimized or the surface presents inherent textures comparable to the tested one.

References

1. Sang-Chan K, Hak Kyeong K, Chae-Gue L, Sang-Bong K (2006) A vision system for identifying structural vibration in civil engineering constructions. In: SICE-ICASE international joint conference 2006, Bexco, 18–21 Oct 2006
2. Yoshio F, Feng MQ, Masanobu S (2009) Cost-effective vision-based system for monitoring dynamic response of civil engineering structures. Struct Control Health Monit. doi:10.1002/stc.360
3. Olaszek P (1999) Investigation of the dynamic characteristic of bridge structures using a computer vision method. Measurement 25:227–236
4. Giorgio B, Alfredo C, Paolo M, Emanuele Z, Maurizio F (2012) Cameras as displacement sensors to get the dynamic motion of a bridge: performance evaluation against traditional approaches. 6th international conference on bridge maintenance, safety and management (IABMAS 2012), Lake Como, 8–12 July 2012
5. Mazen Wahbeh A, Caffrey JP, Masri SF (2003) A vision-based approach for the direct measurement of displacements in vibrating systems. Smart Mater Struct 12:785–794

6. Caetano E, Silva S, Bateira J, Caetano E, Silva S, Bateira J (2011) A vision system for vibration monitoring of civil engineering structures. Exp Tech 35:74–82
7. Giorgio B, Alfredo C, Emanuele Z (2012) Dynamic characterization of flexible structures through vision-based vibration measurements. In: Proceedings of the international modal analysis conference IMAC XXX, Jacksonville, 30 Jan–2 Feb 2012
8. Waterfall PW, Mcdonald JHG, McCormick NJ (2012) Targetless precision monitoring of road and rail bridges using video cameras. In: Proceedings of the 6th international conference on bridge maintenance, safety and management, Stresa
9. Jong Jae L, Jong Jae L, Jong Jae L, Jong Jae L, Masanobu S, Masanobu S (2006) A vision-based system for remote sensing of bridge displacement. NTD E Int 39(2066):425–431
10. Piotr K, Piotr K, Piotr K, Piotr K (2009) application of modal analysis supported by 3D vision-based measurements. J Theor Appl Mech 47 (4):855–870, Warsaw
11. Cantatore A, Cigada A, Sala R, Zappa E (2009) Hyperbolic tangent algorithm for periodic effect cancellation in sub-pixel resolution edge displacement measurement. Measurement 42:1226–1232
12. Cigada SZ (2003) On the reduction of subpixel error in image-based displacement sensor. In: XVII IMEKO World Congress, Dubrownik, 22–27 June 2003
13. Giorgio B, Alfredo C, Marcello V, Emanuele Z (2011) Vision-based vibration monitoring of a large steel structure, Experimental Vibration Analysis for Civil Engineering Structures (EVACES), Varenna, 3–5 Oct 2011

Chapter 10
Using an Instrumented Tractor-Trailer to Detect Damage in Bridges

Eugene J. OBrien and Jennifer Keenahan

Abstract This paper investigates an alternative to Structural Health Monitoring (SHM) which involves no sensors on the bridge itself. It uses a vehicle fitted with accelerometers on its axles as a method of monitoring the dynamic behavior of the bridge, which in turn gives an indication of the bridge's structural condition. The concept, known as 'drive by' bridge inspection, may be of particular value after an extreme event, such as an earthquake or a flood, where a rapid indication of bridge condition is needed. In the paper, a two dimensional numerical model is described of a 3-axle truck towing a half-car trailer. The vehicle-bridge dynamic interaction is modeled to test the effectiveness of the approach in identifying the structural damping of the bridge. The damping of the bridge is used here as an indicator of damage. The accelerations in the two axles of the trailer are subtracted to remove the effect of the road profile. Results indicate that the algorithm works well and is not sensitive to transverse position of the vehicle on the bridge.

Keywords Bridge monitoring • Bridges • Damage detection models

10.1 Introduction

The task of detecting damage in bridges traditionally consists of visual inspections. These however are labor intensive and are often an unreliable way of determining the true condition. Recently there has been a move towards sensor based analysis of bridge condition. Existing monitoring techniques involve the direct instrumentation of the structure – commonly referred to as Structural Health Monitoring (SHM) [1–3]. More recently, a small number of authors have shifted to the instrumentation of a vehicle, rather than the bridge, which can be less expensive and less time-consuming. This approach is referred to as 'drive-by' bridge inspection [4]. The feasibility of detecting frequencies from the dynamic response of an instrumented vehicle passing over a bridge has been verified theoretically by Yang et al. [5], in field trials [6, 7] and in laboratory investigations [4, 8–10]. As an alternative to detecting changes in frequency, Yabe and Miyamoto [10] use the mean displacement of the rear axle of a city bus passing over a bridge a large number of times as a damage indicator. Kim et al. [12] construct scaled Vehicle Bridge Interaction (VBI) laboratory experiments and consider the use of autoregressive coefficients as a damage indicator. The analysis of damping has been considered to a lesser extent [13]. However, recent evidence suggests that damping is quite sensitive to damage in structural elements and in some cases, more sensitive than other indicators [14, 15].

This paper describes a novel approach that uses a truck-trailer vehicle system, fitted with accelerometers on the trailer axles, to detect changes in the damping of a bridge which would indicate deterioration of the bridge's condition. The concept is that the relatively heavy truck dynamically excites the bridge while sensors in the trailer are used to monitor the resulting vibrations. For numerical simulations, a VBI model is created in Matlab. The trailer axles are assigned identical properties – as can easily be the case with a simple trailer. The axle accelerations from the front and rear axles of the trailer are *subtracted* from one another. Each trailer axle is excited by the same road profile and by an element of bridge vibration at a different

E.J. OBrien (✉) • J. Keenahan
School of Civil, Structural and Environmental Engineering, University College Dublin,
Newstead Building, Belfield Campus, Dublin 4, Ireland
e-mail: eugene.obrien@ucd.ie

A. Cunha (ed.), *Topics in Dynamics of Bridges, Volume 3: Proceedings of the 31st IMAC, A Conference on Structural Dynamics, 2013*, Conference Proceedings of the Society for Experimental Mechanics Series 38,
DOI 10.1007/978-1-4614-6519-5_10, © The Society for Experimental Mechanics, Inc. 2013

point in time. Subtracting the signals, time shifted by the interval between axle arrivals, has the effect of removing most of the influence of the road profile. This is a key feature of this approach and is the reason why the results are better than in simpler drive-by monitoring concepts. The vehicle is simulated crossing different paths through the road profile to assess sensitivity to transverse position of the vehicle on the bridge.

10.2 Vehicle-Bridge Interaction Model

The truck-trailer model can be seen in Fig. 10.1. The truck is a three axle, five-degree-of-freedom rigid vehicle. The five degrees-of-freedom account for the axle hop displacements of each of the three axles, $y_{u,i}$ $(i = 1, 2, 3)$, sprung mass bounce displacement, $y_{s,1}$, and sprung mass pitch rotation, $\theta_{s,1}$. The body of the vehicle is represented by the sprung mass, $m_{s,1}$, and the axle components are represented by the unsprung masses, $m_{u,1}$, $m_{u,2}$ and $m_{u,3}$ respectively. The axle masses connect to the road surface via springs of stiffness $K_{t,1}$, $K_{t,2}$ and $K_{t,3}$, while the body mass is connected to the tires by springs of stiffness $K_{s,1}$, $K_{s,2}$ and $K_{s,3}$ with viscous dampers of value $C_{s,1}$, $C_{s,2}$ and $C_{s,3}$. This combination represents the suspension of the truck system.

The trailer is a two axle, four-degree-of-freedom half-car suspension model. The four degrees-of-freedom account for axle hop displacements of each of the two axles, $y_{u,i}$ $(i = 4, 5)$, sprung mass bounce displacement, $y_{s,2}$ and sprung mass pitch rotation, $\theta_{s,2}$. The body of the vehicle is represented by the sprung mass, $m_{s,2}$, and the axle components are represented by the unsprung masses, $m_{u,4}$ and $m_{u,5}$. The suspension springs have stiffness $K_{t,4}$ and $K_{t,5}$, while the tires springs have stiffness $K_{s,4}$ and $K_{s,5}$. The viscous dampers have coefficients, $C_{s,4}$ and $C_{s,5}$. Tire damping is assumed to be negligible here for both the tractor and trailer and is thus omitted. The model also accounts for the sprung mass moments of inertia $I_{s,1}$ and $I_{s,2}$ for the truck and trailer respectively. The center of gravity of the truck is taken to be at two thirds the wheel base length from the front axle, and the center of gravity of the trailer is taken to be central between the axles. The truck and trailer vehicle properties are gathered from the literature [16–18] and presented in Table 10.1. The geometry and mass of the truck are obtained from a manufacturer specification for a 30 t three-axle truck [19].

The equations of motion of the vehicle are obtained by imposing equilibrium of all forces and moments acting on the vehicle and expressing them in terms of the degrees of freedom. They are given by

$$\mathbf{M}_v \ddot{\mathbf{y}}_v + \mathbf{C}_v \dot{\mathbf{y}}_v + \mathbf{K}_v \mathbf{y}_v = \mathbf{f}_v \tag{10.1}$$

where $\mathbf{M}_v$, $\mathbf{C}_v$ and $\mathbf{K}_v$ are the mass, damping and stiffness matrices of the vehicle respectively. The $(n \times 1)$ vectors $\mathbf{y}_v$, $\dot{\mathbf{y}}_v$ and $\ddot{\mathbf{y}}_v$ contain the of vehicle displacements, their velocities and accelerations respectively. The vector $\mathbf{f}_v$ contains the time varying interaction forces applied by the vehicle to the bridge:

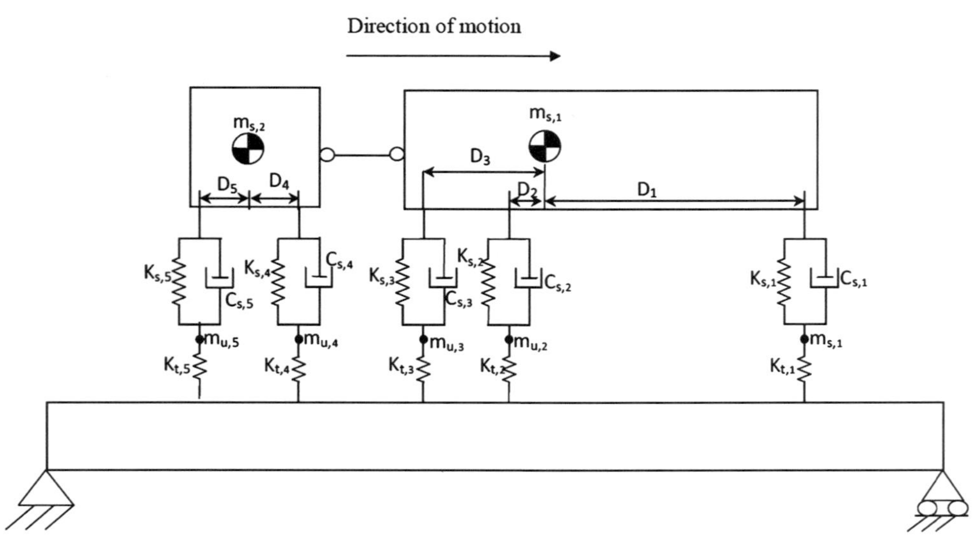

Fig. 10.1 Truck-trailer model

Table 10.1 Truck and trailer properties

Property	Unit	Truck symbol	Truck value	Trailer symbol	Trailer value
Body mass	kg	$m_{s,1}$	27,100	$m_{s,2}$	400
Axle masses	kg	$m_{u,1}$	700	$m_{u,4}$	50
		$m_{u,2}$	1,100	$m_{u,5}$	50
		$m_{u,3}$	1,100		
Suspension stiffness	N m^{-1}	$K_{s,1}$	4×10^5	$K_{s,4}$	4×10^5
		$K_{s,2}$	1×10^6	$K_{s,5}$	4×10^5
		$K_{s,3}$	1×10^6		
Suspension damping	Ns m^{-1}	$C_{s,1}$	10×10^3	$C_{s,4}$	10×10^3
		$C_{s,2}$	20×10^3	$C_{s,5}$	10×10^3
		$C_{s,3}$	20×10^3		
Tire stiffness	N m^{-1}	$K_{t,1}$	1.75×10^6	$K_{t,4}$	1.75×10^6
		$K_{t,2}$	3.5×10^6	$K_{t,5}$	1.75×10^6
		$K_{t,3}$	3.5×10^6		
Moment of inertia	kg m^2	$I_{s,1}$	1.56×10^5	$I_{s,2}$	241.67
Distance of axle to center of gravity	m	D_1	4.57	D_4	1
		D_2	1.43	D_5	1
		D_3	3.23		
Body mass frequency	Hz	$f_{body,1}$	1.12	$f_{body,2}$	1.77
Axle mass frequency	Hz	$f_{axle,1}$	8.84	$f_{axle,4}$	33.1
		$f_{axle,2}$	10.18	$f_{axle,5}$	33.1
		$f_{axle,3}$	10.22		

$$\mathbf{f}_{\mathbf{v}} = \{0 \quad 0 - F_{t,1} - F_{t,2} - F_{t,3} \quad 0 \quad 0 - F_{t,4} - F_{t,5}\}^{\mathrm{T}} \tag{10.2}$$

The term $F_{t,i}$ represents the dynamic interaction force at wheel i given by (10.3).

$$F_{t,i} = K_{t,i}(y_{u,i} - y_{br,i} - r_i); \tag{10.3}$$

The equations of motion of the VBI model are shown below. The nine degrees of freedom correspond to body bounce of the truck (10.4) and trailer (10.5), body pitch of the truck (10.6) and trailer (10.7), and axle hop for each of the five axles; the latter can be represented by the form given in (10.8). For $i = 1, 4$, $D_i\dot{\theta}_{s,j}$ is taken as a positive number, and for $i = 2, 3, 5$, $D_i\dot{\theta}_{s,j}$ is taken as negative.

$$\begin{aligned} m_{s,1}\ddot{y}_{s,1} &+ C_{s,1}(\dot{y}_{s,1} + D_1\dot{\theta}_{s,1} - \dot{y}_{u,1}) + K_{s,1}(y_{s,1} + D_1\theta_{s,1} - y_{u,1}) + C_{s,2}(\dot{y}_{s,1} - D_2\dot{\theta}_{s,1} - \dot{y}_{u,2}) \\ &+ K_{s,2}(y_{s,1} - D_2\theta_{s,1} - y_{u,2}) + C_{s,3}(\dot{y}_{s,1} - D_3\dot{\theta}_{s,1} - \dot{y}_{u,3}) + K_{s,3}(y_{s,1} - D_3\theta_{s,1} - y_{u,3}) = 0 \end{aligned} \tag{10.4}$$

$$\begin{aligned} m_{s,2}\ddot{y}_{s,2} &+ C_{s,4}(\dot{y}_{s,2} + D_4\dot{\theta}_{s,2} - \dot{y}_{u,4}) + K_{s,4}(y_{s,2} + D_4\theta_{s,2} - y_{u,4}) \\ &+ C_{s,5}(\dot{y}_{s,2} - D_5\dot{\theta}_{s,2} - \dot{y}_{u,5}) + K_{s,5}(y_{s,2} - D_5\theta_{s,2} - y_{u,5}) = 0 \end{aligned} \tag{10.5}$$

$$\begin{aligned} I_{s,1}\ddot{\theta}_{s,1} &+ D_1\left[C_{s,1}(\dot{y}_{s,1} + D_1\dot{\theta}_{s,1} - \dot{y}_{u,1}) + K_{s,1}(y_{s,1} + D_1\theta_{s,1} - y_{u,1})\right] \\ &- D_2\left[C_{s,2}(\dot{y}_{s,1} - D_2\dot{\theta}_{s,1} - \dot{y}_{u,2}) + K_{s,2}(y_{s,1} - D_2\theta_{s,1} - y_{u,2})\right] \\ &- D_3\left[C_{s,3}(\dot{y}_{s,1} - D_3\dot{\theta}_{s,1} - \dot{y}_{u,3}) + K_{s,3}(y_{s,1} - D_3\theta_{s,1} - y_{u,3})\right] = 0 \end{aligned} \tag{10.6}$$

$$\begin{aligned} I_{s,2}\ddot{\theta}_{s,2} &+ D_4\left[C_{s,4}(\dot{y}_{s,2} + D_4\dot{\theta}_{s,2} - \dot{y}_{u,4}) + K_{s,4}(y_{s,2} + D_4\theta_{s,2} - y_{u,4})\right] \\ &- D_5\left[C_{s,5}(\dot{y}_{s,2} - D_5\dot{\theta}_{s,2} - \dot{y}_{u,5}) + K_{s,5}(y_{s,2} - D_5\theta_{s,2} - y_{u,5})\right] = 0 \end{aligned} \tag{10.7}$$

$$m_{u,i}\ddot{y}_{u,i} - C_{s,i}\left(\dot{y}_{s,j} \pm D_i\dot{\theta}_{s,j} - \dot{y}_{u,i}\right) - K_{s,1}\left(y_{s,j} \pm D_i\theta_{s,j} - y_{u,i}\right) - P_i + F_{t,i} = 0; \qquad i = 1, 2, \ldots, 5; j = 1, 2 \tag{10.8}$$

The bridge model used here is a simply supported 15 m Finite Element beam that consists of twenty discretized beam elements with four degrees of freedom. The beam therefore has a total of $n = 42$ degrees of freedom. It has a constant modulus of elasticity $E = 3.5 \times 10^{10}$ N m^{-2}, mass per unit length, $\mu = 28\ 125$ kg m^{-1} and second moment of area, $J = 0.5273$ m^4. The first natural frequency of the beam is 5.65 Hz. The response of a discretized beam model to a series of moving time-varying forces is given by the system of equations:

$$\mathbf{M_b}\ddot{\mathbf{y}}_\mathbf{b} + \mathbf{C_b}\dot{\mathbf{y}}_\mathbf{b} + \mathbf{K_b}\mathbf{y_b} = \mathbf{N_b}\mathbf{f_{int}} \tag{10.9}$$

where $\mathbf{M_b}$, $\mathbf{C_b}$ and $\mathbf{K_b}$ are the $(n \times n)$ global mass, damping and stiffness matrices of the beam model respectively and $\mathbf{y_b}$, $\dot{\mathbf{y}}_\mathbf{b}$ and $\ddot{\mathbf{y}}_\mathbf{b}$ are the $(n \times 1)$ global vectors of nodal bridge displacements and rotations, their velocities and accelerations respectively. The product $\mathbf{N_b f_{int}}$ is the $(n \times 1)$ global vector of forces applied to the bridge nodes. The vector $\mathbf{f_{int}}$ contains the interaction forces between the vehicle and the bridge and is described using the following vector:

$$\mathbf{f_{int}} = \mathbf{P} + \mathbf{F_t} \tag{10.10}$$

where $\mathbf{P}$ is the static axle load vector and $\mathbf{F_t}$ contains the dynamic wheel contact forces of each axle. The matrix $\mathbf{N_b}$ is a $(n \times n_f)$ location matrix that distributes the n_f applied interaction forces on beam elements to equivalent forces acting on nodes. This location matrix can be used to calculated bridge displacement under each wheel, $\mathbf{y_{br}}$:

$$\mathbf{y_{br}} = \mathbf{N_b^T}\ \mathbf{y_b} \tag{10.11}$$

The damping ratio of the bridge, ξ, is varied in simulations to assess the system's potential as an indicator of changes in damping. Although complex damping mechanisms may be present in the structure, viscous damping is typically used for bridge structures and is deemed to be sufficient to reproduce the bridge response accurately. Therefore, Rayleigh damping is adopted here to model viscous damping:

$$\mathbf{C_b} = \alpha\mathbf{M_b} + \beta\mathbf{K_b} \tag{10.12}$$

where α and β are constants. The damping ratio is assumed to be the same for the first two modes [20] and α and β are obtained from $\alpha = 2\ \xi\ \omega_1\omega_2/(\omega_1 + \omega_2)$ and $\beta = 2\ \xi/(\omega_1 + \omega_2)$ where ω_1 and ω_2 are the first two natural frequencies of the bridge [21].

The dynamic interaction between the vehicle and the bridge is implemented in Matlab. The vehicle and the bridge are coupled at the tire contact points via the interaction force vector, $\mathbf{f_{int}}$. Combining (10.1) and (10.9), the coupled equation of motion is formed as

$$\mathbf{M_g}\ddot{\mathbf{u}} + \mathbf{C_g}\dot{\mathbf{u}} + \mathbf{K_g}\mathbf{u} = \mathbf{F} \tag{10.13}$$

where $\mathbf{M_g}$ and $\mathbf{C_g}$ are the combined system mass and damping matrices respectively, $\mathbf{K_g}$ is the coupled time-varying system stiffness matrix and $\mathbf{F}$ is the system force vector. The vector, $\mathbf{u} = \{\mathbf{y_v},\ \mathbf{y_b}\}^\mathbf{T}$ is the displacement vector of the system. The equations for the coupled system are solved using the Wilson-Theta integration scheme [22, 23]. The optimal value of the parameter $\theta = 1.420815$ is used for unconditional stability in the integration schemes [24]. The scanning frequency used for all simulations is 1,000 Hz.

10.3 Concept of Subtracting Axle Accelerations to Detect Bridge Damping

The vehicle is simulated traveling over a 100 m approach length followed by a 15 m simply supported bridge at 20 m s^{-1}. The Class 'A' road profile is generated according to the ISO standard [25]. This is extrapolated into a number of different paths, where each path is correlated with those adjacent to it. Figure 10.2 illustrates two paths through this road profile. The vehicle is simulated crossing each path separately to assess the sensitivity of the algorithm to transverse position. This is repeated six times, once for each level of damping (from 0% to 5%), representing different degrees of damage.

The trailer vertical accelerations are transformed from the time domain into the frequency domain using the Fast Fourier Transform. Plots of Power Spectral Density (PSD) against frequency can be seen in Fig. 10.3. In the PSD for an individual

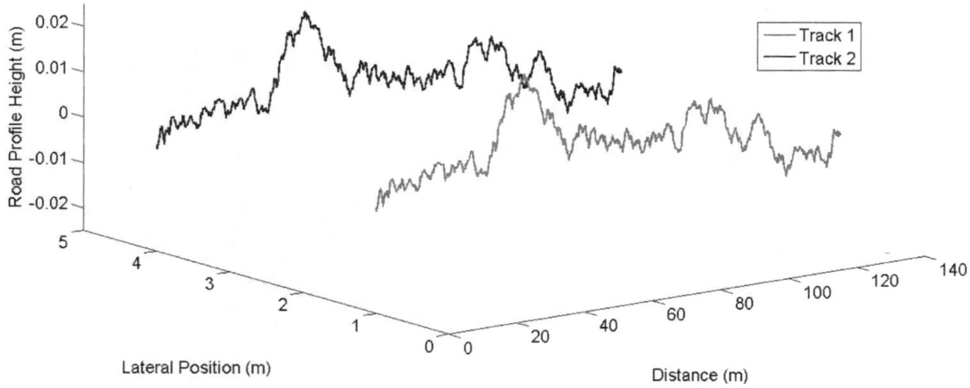

Fig. 10.2 Class 'A' road profile [25]

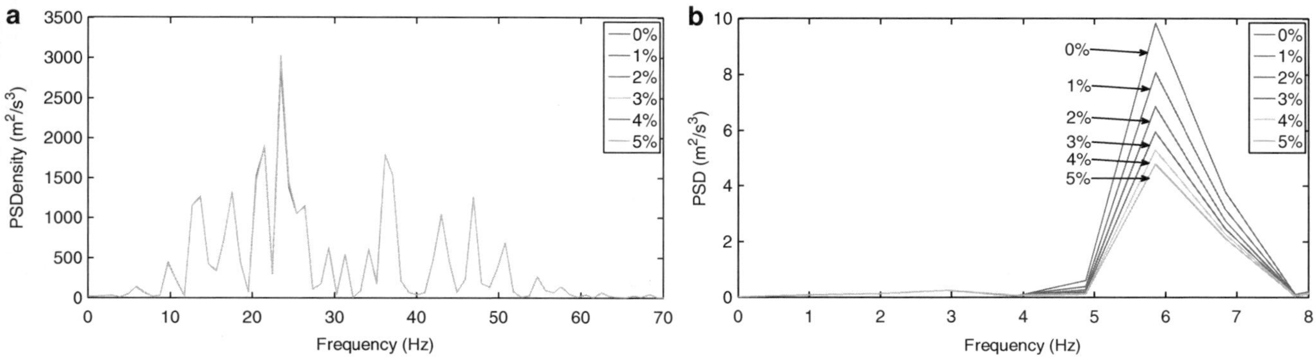

Fig. 10.3 PSD of accelerations for vehicle traveling along path 1 at 20 m s^{-1} over a 15 m bridge; (**a**) Trailer axle 1 accelerations (**b**) Axle acceleration difference

axle of the trailer (Fig. 10.3a), there is no peak corresponding to the bridge frequency (5.65 Hz) and there is no clear distinction between the different levels of damping (all six plots are on top of one another). The vibration of the vehicle dominates each spectrum. This is because the ratio of height of road irregularities to bridge displacements is too large for the bridge to have a significant influence on the vehicle.

However, when the trailer axle accelerations are subtracted from one another, allowing for the time shift, clear peaks become visible corresponding to the first natural frequency of the bridge, seen in Fig. 10.3b. The first trailer axle is excited by the road profile and the bridge displacements as it passes each point on the bridge. The second trailer axle is excited by the same road profile and bridge displacements at different instants in time. The differences between the excitations include no element of road profile, only consisting of time lagged differences in bridge response. As such, this difference plot can be used to identify the influence of the bridge alone, without 'contamination' from the excitations due to road surface profile. It follows that the effect of bridge damping, hardly visible in Fig. 10.3a, is clearly evident in Fig. 10.3b. It is apparent from Fig. 10.3b that the magnitude of the peak decreases for higher levels of damping. This suggests that a truck-trailer vehicle system has the potential to be a practical method of detecting changes in PSD which then may be used as an indicator of changes in bridge damping. This same trend can be seen in Fig. 10.4a, b.

This illustrates that the method presented here is not sensitive to transverse position of the vehicle on the bridge. What can be noted, is that the magnitude of Power Spectral Density is larger for the second path (Fig. 10.4b) compared with the first path (Fig. 10.3b). This is due to the differing levels of excitation of the vehicle.

The PSD of the bridge mid span accelerations, which is the acceleration reading from instrumentation of the bridge as opposed to the vehicle, are also found. A peak occurs at 5.85 Hz here also. *This suggests that instrumentation of the vehicle can be of similar accuracy to results found by instrumenting the bridge.* There is a small difference between the frequencies that were predicted (5.65 Hz) and where the peak occurs (5.85 Hz) in Figs. 10.3 and 10.4. The inaccuracy appears to be due to the spectral resolution (± 0.48 Hz), which can be improved by driving the vehicle at a slower speed.

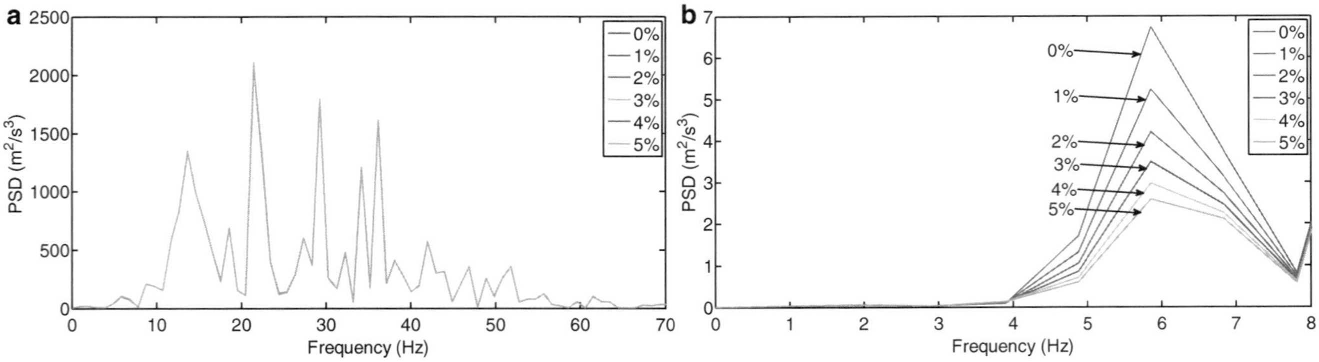

Fig. 10.4 PSD of accelerations for vehicle travelling along path 2 at 20 m s^{-1} over a 15 m bridge; (**a**) Trailer axle 1 accelerations (**b**) Axle acceleration difference

10.4 Conclusions

This paper investigates the feasibility of using an instrumented truck-trailer vehicle model to monitor damping in a bridge. A method is presented that involves the subtraction of axle accelerations to remove much of the influence of the road profile. The results indicate that bridge frequency and changes in damping can be detected when the axle accelerations of the trailer are subtracted from one another. Results for the drive-by system are of similar quality to results for an accelerometer located on the bridge and the algorithm appears to be insensitive to transverse position of the vehicle on the road.

Acknowledgements The authors wish to express their gratitude for the financial support received from Science Foundation Ireland towards this investigation under the US-Ireland Research Partnership Scheme.

References

1. Brownjohn JMW (2007) Structural health monitoring of civil infrastructure. Philos Trans R Soc A Math Phys Eng Sci 365:589–622
2. Chang PC, Flatau A, Liu SC (2003) Review paper: health monitoring of civil infrastructure. Struct Health Monit 2:257–267
3. Farrar CR, Worden K (2007) An introduction to structural health monitoring. Philos Trans R Soc A Math Phys Eng Sci 365:303–315
4. Kim CW, Kawatani M (2009) Challenge for a drive-by bridge inspection. In: Proceedings of the tenth international conference on structural safety and reliability, Osaka, 2009, pp 758–765
5. Yang YB, Lin CW, Yau JD (2004) Extracting bridge frequencies from the dynamic response of a passing vehicle. J Sound Vib 272:471–493
6. Lin CW, Yang YB (2005) Use of a passing vehicle to scan the fundamental bridge frequencies: an experimental verification. Eng Struct 27:1865–1878
7. Yang YB, Chang KC (2009) Extraction of bridge frequencies from the dynamic response of a passing vehicle enhanced by the EMD technique. J Sound Vib 322:718–739
8. Oshima Y, Yamaguchi T, Kobayashi Y, Sugiura K (2008) Eigenfrequency estimation for bridges using the response of a passing vehicle with excitation system. In: Proceedings of the fourth international conference on bridge maintenance, safety and management, Seoul, pp 3030–3037
9. Toshinami T, Kawatani M, Kim CW (2010) Feasibility investigation for identifying bridge's fundamental frequencies from vehicle vibrations. In: Proceedings of the fifth international conference on bridge maintenance, safety and management, Philadelphia, pp 317–322
10. Cerda F, Garrett J, Bielak J, Barrera J, Zhuang Z, Chen S, McCann M, Kovačević J (2012) Indirect structural health monitoring in bridges: scale experiments. In: Proceedings of the sixth international conference for bridge maintenance and safety, Stresa, pp 346–353
11. Yabe A, Miyamoto A (2012) Bridge condition assessment for short and medium span bridges by vibration responses of city bus. In: Proceedings of the sixth international conference for bridge maintenance and safety, Stresa, pp 195–202
12. Kim CW, Isemoto R, Sugiura K, Kawatani M (2012) Structural diagnosis of bridges using traffic-induced vibration measurements. In: Proceedings of the sixth international conference for bridge maintenance and safety, Stresa, pp 423–430
13. Williams C, Salawu OS (1997) Damping as a damage indication parameter. In: Proceedings of the fifteenth international modal analysis conference, Orlando, pp 1531–1536
14. Curadelli RO, Riera JD, Ambrosini D, Amani MG (2008) Damage detection by means of structural damping identification. Eng Struct 30:3497–3504
15. Modena C, Sonda D, Zonta D (1999) Damage localization in reinforced concrete structures by using damping measurements. Key Eng Mater 167:132–141
16. McGetrick PJ, González A, OBrien EJ (2009) Theoretical investigation of the use of a moving vehicle to identify bridge dynamic parameters. Insight Non-Destr Test Cond Monit 51:433–438
17. OBrien EJ, Li Y, González A (2006) Bridge roughness index as an indicator of bridge dynamic amplification. Comput Struct 84:759–769

18. Zhu XQ, Law SS (2001) Orthogonal function in moving loads identification in a multi-span bridge. J Sound Vib 245:329–345
19. DAF Trucks Limited, FAT CF75 30 t Specification sheet, 2012
20. Yang YB, Yau JD, Wu YS (2004) Vehicle-bridge interaction dynamics: with applications to high-speed railways. World Scientific Publishing, Singapore
21. Clough RW, Penzien J (1975) Dynamics of structures. McGraw-Hill, New York
22. Bathe KJ, Wilson EL (1976) Numerical methods in finite element analysis. Prentice-Hall, Englewood Cliffs
23. Tedesco JW, McDougal WG, Ross CA (1999) Structural dynamics: theory and applications. Addison Wesley Longman, Menlo Park
24. Weaver W, Johnston PR (1987) Structural dynamics by finite elements. Prentice-Hall, Englewood Cliffs
25. International Organisation for Standardisation ISO 8608, Mechanical vibration-road surface profiles - reporting of measured data, 1995

Chapter 11
Sensitivity Analysis of the Lateral Damping of Bridges for Low Levels of Vibration

A.R. Ortiz, C.E. Ventura, and S.S. Catacoli

Abstract Damping is a dynamic property that has not been fully understood. In particular, damping in bridges requires further study in order to gain a better understanding of the parameters that define its value and variation. In practice a constant ratio of 5 % of the critical damping is commonly used in seismic analysis and design. However, this may not be accurate for different levels of shaking. In addition, the experimental methods to estimate damping can provide different damping ratios for the same structure, which increases the uncertainties on the damping estimates. This paper presents a sensitivity analysis of the damping in bridges for different types of excitation and low levels of vibration. Data collected from different sources is used to conduct the assessment, namely, scaled model tests, ambient vibration tests, and seismic ground motions from instrumented bridges in California. A discussion of the results using several frequency and time domain methods for damping estimation is presented.

Keywords Lateral damping • Bridges • Low level shaking • Sensitivity analysis • Modal identification

11.1 Introduction

Damping is an important structural characteristic that modifies the structural response to dynamic excitation. In structural engineering, damping plays an important role in the assessment of the structural demands due to wind and seismic loads that historically had produced significant human and economic losses, as well as to service loads produced by humans or machines.

An appropriate estimation of the damping ratio is connected to the correct assessment of the structural response. Traditionally, bridge engineers have used a damping ratio of 5 % as representative of the lateral damping of standard bridges under seismic loads. However, further studies are required to understand the sensitivity of damping ratio to different parameters, such as: types and levels of ground motion, bridge's structural configuration and structural element's materials [1]. In addition, the use passive devices as viscous dampers to improve the seismic performance of bridges has increased the need to gain a better understanding of the damping ratio in bridges.

This paper initially presents the experimental results of a 3-DOF's lumped mass model used to explore the relation between damping and the frequency and amplitude of the external loads. Then, data from bridges included in the ambient vibration database of the Earthquake Engineering Research Facility (EERF) at the University of British Columbia are used to estimate the damping ratio of bridges for low levels of vibration. Finally, the response of instrumented bridges in California is used to calculate the damping ratio, especially for low levels of shaking.

A.R. Ortiz (✉)
Universidad de Medellin, Civil Engineering Department, Medellin, Colombia
e-mail: arortiz@udem.edu.co

C.E. Ventura • S.S. Catacoli
University of British Columbia, Civil Engineering Department, Vancouver, Canada
e-mail: ventura@civil.ubc.ca; sesamory@interchange.ubc.ca

A. Cunha (ed.), *Topics in Dynamics of Bridges, Volume 3: Proceedings of the 31st IMAC, A Conference on Structural Dynamics, 2013*, Conference Proceedings of the Society for Experimental Mechanics Series 38,
DOI 10.1007/978-1-4614-6519-5_11, © The Society for Experimental Mechanics, Inc. 2013

11.2 Damping Estimation Methods

11.2.1 *Free Vibration Tests*

This approach uses the measurement of the free structural response after the application of an impulsive load or an imposed displacement to estimate the damping ratio [2]. The measured decay in the response is assumed to fit an exponential function from which the damping ratio is directly estimated (11.1). For Multi-Degrees of Freedom Systems (MDOF's) is important to ensure that the response analyzed only corresponds to the mode of vibration of interest. In this paper, the data that correspond to the lumped mass model was filtered to capture only the first mode of vibration.

$$\xi = \frac{1}{2\pi m} \ln\left(\frac{x_i}{x_{m+1}}\right) \tag{11.1}$$

Where m is the successive numbers of peaks in the damping estimation zone, x_i is the amplitude for peak i and x_{m+1} is the amplitude for peak $m + 1$.

11.2.2 *Forced Vibration Tests*

Ambient Vibration Testing (AVT) can be used to estimate the damping ratio of structures. AVT involves measuring a structure's response to typical forces that it is subjected to every day. These ambient forces can be wind, traffic, human activities, etc. This method of testing provides a cheap, non-invasive, and non-destructive method for obtaining modal parameters of structures. The response obtained from these tests is characteristic of the true operating conditions of the structure. To obtain the modal parameters of the structure, Modal Operational Analysis algorithms are used to process the data. The Enhanced Frequency Domain Decomposition Method (EFDD) and the Stochastic Subspace Identification Method (SSI) are commonly used, further information about these algorithms is provided in Magalhaes et al. [3].

Harmonic excitations and frequency response functions (FRF) are also used to estimate damping ratios. In this case, the identification uses the input/output relationships of the system and constructs a mathematical model in frequency domain that can replicate that behavior. Once FRF's from each of the system inputs to each of the outputs is determined, each FRF is modeled as a ratio of two polynomials in the Laplace variable and then used to form a state space representation for the structural system from which the damping ratio is obtained [4].

In addition to the methods described above, this paper considers the force vibration data imposed by the ground shaking undergone for a set of instrumented bridges in California. This approach is considered particularly useful to determine the damping ratio for the transverse mode of vibration, as ground motions provide energy to excite the lateral modes of vibrations. The EFDD and SSI methods were used to estimate the damping ratios.

11.3 Description of the Tests

11.3.1 *Laboratory Tests*

A simple 3-DOF's lumped mass model was evaluated in order to characterize the dynamic behavior. The test aims to capture in a simple model the influence of amplitude and frequency of the applied external load on the estimated damping ratio. The model consists of a 1 m aluminum bar with three lumped masses uniformly distributed. The first frequency of vibration is tuned to be 0.99 Hz. The measuring equipment consists of four wireless MEMs tri-axial accelerometers placed at each of the masses and at the base of the shaker. These are light accelerometers which response is linear for low frequencies, considered adequate for this type of test (Fig. 11.1).

The lumped mass model was subjected to free vibration tests with different amplitudes of the initial imposed displacement. In addition, a small shaker was used to apply sinusoidal sweeps with constant frequency and amplitude. Frequency and amplitude were varied to consider a range of frequencies from 0.1 to 10 Hz, and amplitudes of 30 %, 60 % and 100 % of the maximum amplitude that the shaker allowed to apply at each frequency.

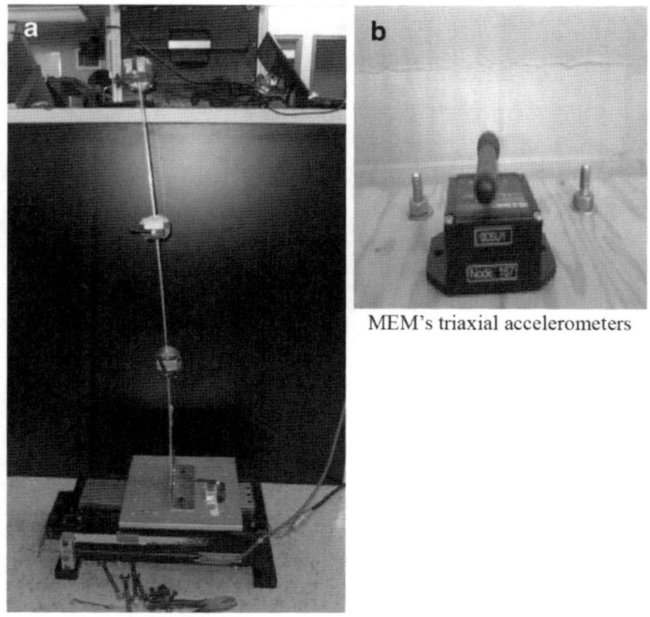

Forced vibration tests

MEM's triaxial accelerometers

Fig. 11.1 3-DOF's lumped mass model test

11.3.2 Ambient Vibration Test

The data from ambient vibration tests performed to ten bridges located in the Province of British Columbia, Canada, are used to estimate the damping ratio of bridges in the transverse direction for low levels of vibration. The considered bridges are classified as short to medium single and multi-span structures, with integral or seat type abutment and built with standard materials. The Table 11.1 summarizes the characteristic of the bridges considered.

11.3.3 Instrumented Bridges

The records from ten instrumented bridges located in California were used to estimate the damping ratio in the transverse direction. The structures analyzed are standard multi-span bridges subjected to different levels of shaking. A description of the bridges and the characteristics of the 24 ground motions recorded are provided in Appendices 1 and 2. Further information is given in Fraino et al. [5].

11.4 Results

11.4.1 Experimental Lumped Mass Model

An exponential decay of the amplitude–typical property of underdamped systems– was observed during the free vibration tests of the lumped mass model (Fig. 11.2). The damping ratios were calculated based on the logarithmic decrement of three consecutive peaks. On the other hand, the damping ratios estimated based on the Frequency Response Function (FRF) were calculated for a fitted above 90 %. The results are presented in Figs. 11.3 and 11.4 as function of average amplitude, which corresponds to the acceleration undergone by the mass at the top. The evaluation indicates a linear increment in the damping ratio as the amplitude of the excitation increases. These results are consistent in the two methods applied.

Table 11.1 Summary of the characteristics of the bridges used for the ambient vibration tests

Structure	Length (m)	Spans No.	Spans Lengths (m)	Width (m)	Clearance (m)	Skew angle (degrees)	Substructure Type	Substructure Abutments	Superstructure type
Duncan bridge	37	1	37	13.4	–	–	–	Integral	Concrete box girder
Royston bridge	43.5	1	43.5	11.0	–	0	–	Integral	Post-tensioned concrete girders
Forbidden bridge	42	1	42	12.8	–	–	–	Integral	Concrete box girder
Dutch bridge	37.5	1	37.5	9.8	–	–	–	Seattype	Steel I-girders
Kensington underpass	60	2	30-30	27	10	45	Multi-column-frames ($\phi = 0.8$ m)	Seattype/ Semi-integral	Continuos-reinforced concrete I-girders
Kanaka creek	41	3	–	20	–	0	Extended Pile Shaft	Seattype	Continuos-reinforced concrete
HWY 24th	48	2	23-23	19	4.9	37	Multi-column-frames ($\phi = 0.8$ m)	Seattype	Continuous – concrete box girders
HWY 10th	71	2	36-36	22.1	9.1	31	Multi-column-frames ($\phi = 1.0$ m)	Seattype	Discontinuous – reinforced concrete I-girders
LHH-EB underpass	135	3	37-58-37	26	5.0	54–57	Multi-column-frames ($\phi = 1.22$ m)	Bent/Seattype	Continuos – concrete steelgirders
Douglas Rd	83	4	12-22-21-21	17	4.6	28	Multi-column-frames ($\phi = 0.6$ m)	Seattype	Disontinuous – concrete box girders

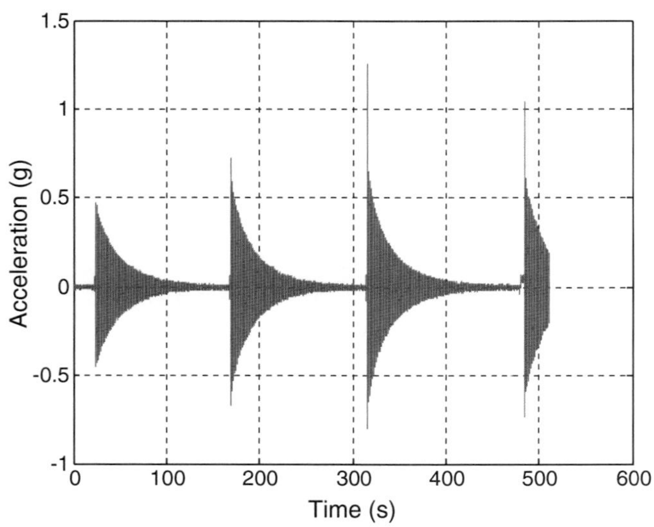

Fig. 11.2 Time record of free vibration tests

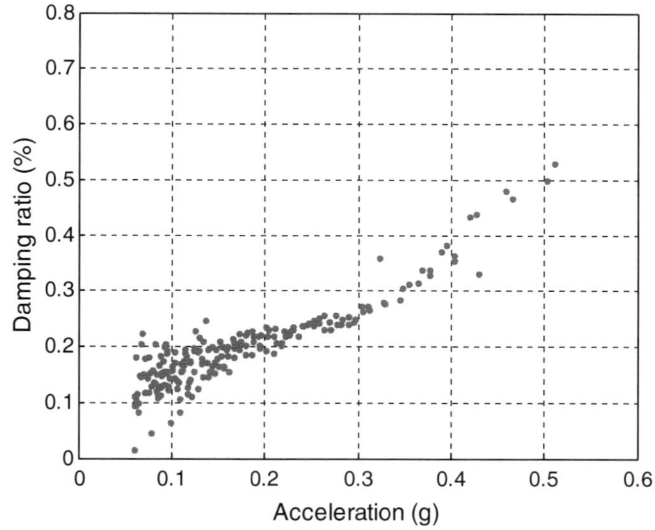

Fig. 11.3 Damping ratios estimated from the logarithmic decrement in free vibration tests

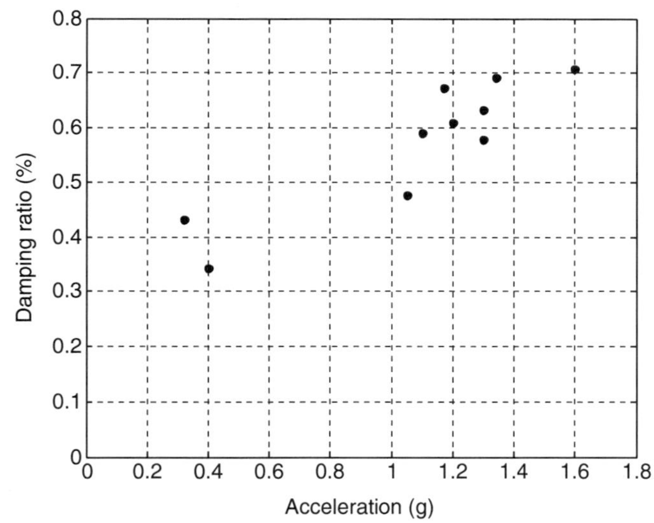

Fig. 11.4 Damping ratios estimated by FRF identification (input–output tests)

Figure 11.5 presents the typical response of a lumped mass model subjected to harmonic excitations at the base. The EFDD method was used to obtain the damping ratios for a set harmonics excitations, with frequencies ranging from 0.1 to 10 Hz. According to the results the damping ratio is independent of the frequency of vibration of the external excitation (Fig. 11.6).

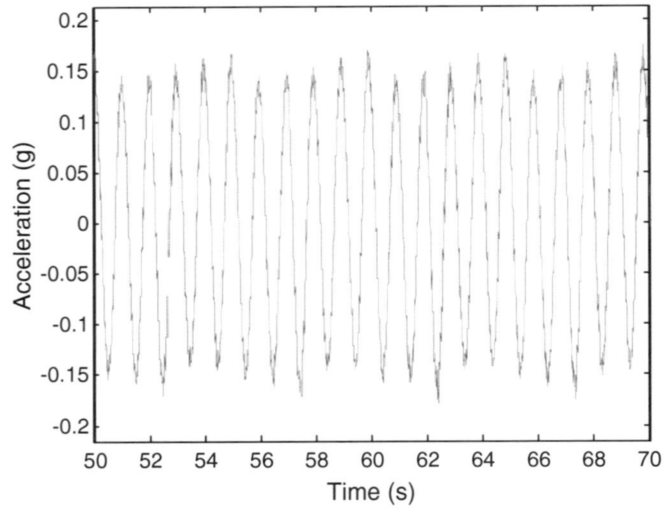

Fig. 11.5 Typical response to harmonic excitation

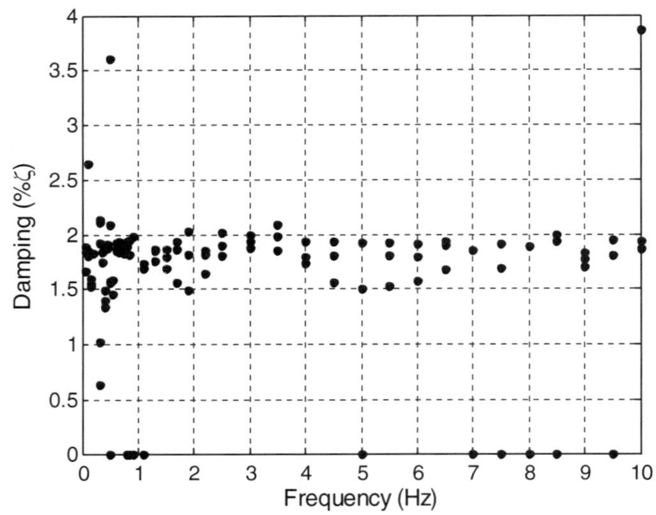

Fig. 11.6 Damping versus frequency of excitation using EFFD method

Table 11.2 Lateral damping ratios obtained using ambient vibration testing

Structure	Frequency (Hz)	Damping ratio transverse mode (%)
Duncan bridge	12.03	1.21
Royston bridge	10.4	1.36
Forbidden bridge	8.73	1.16
Dutch bridge	6.11	1.43
Kensington underpass	3.28	1.30
Kanaka creek	13.84	1.31
HWY 24th	10.66	1.53
HWY 10th	4.67	1.71
LHH-EB underpass	1.72	1.90
Douglas Rd	0.42	0.90

11.4.2 Lateral Damping of Bridges Using Ambient Vibration Tests

The damping ratios of ten bridges considered are shown in Table 11.2. The values reported correspond to damping ratios for the first transverse mode of vibration and are estimated using the EFDD technique. The results suggest that for low levels of shaking–induced by ambient vibrations in this case– the lateral damping ratios of standard short and medium span bridges are below 2 %.

11.4.3 Lateral Damping of Bridges Using Ground Motion Data

The damping ratios of the bridges subjected to different levels of shaking were calculated by using the EFDD and the SSI method provided by Software Artemis Extractor [6]. The results are reported in Table 11.3 as function of the maximum acceleration recorded at the deck, and displayed in Figs. 11.7 and 11.8. For levels of shaking smaller than 0.06 g the EFDD method estimates damping ratios from 1.14 % to 2.21 %. For the same range of amplitudes the SSI method estimates ratios

Table 11.3 Lateral damping ratios obtained using ground motion data

Bridge name	Deck acceleration (g)	Freq. (Hz)	EFDD Damp(%ξ)	SSI Damp(%ξ)
AlvisoOverpassK	0.04	1.661	2.131	4.713
AlvisoOverpassL	0.05	2.568	2.437	4.317
LakeCrowley	0.06	5.342	0.356	0.75
LakeCrowley	0.04	5.239	2.213	2.94
LakeCrowley	0.02	5.621	1.148	2.171
Highway101-156Overpass	0.02	1.598	1.464	5.81
Highway101-156Overpass	0.09	2.093	1.19	2.408
Highway101-156Overpass	0.02	2.414	1.5	1.614
BarrelSprings	0.02	2.844	1.977	4.783
BarrelSprings	0.09	3.003	2.7	3.05
ViaCalifornia	0.02	2.688	1.356	3.98
ViaCalifornia	0.09	2.675	2.675	2.275
ViaCalifornia	0.06	2.623	1.753	3.35
Meloland	0.04	3.023	2.01	2.811

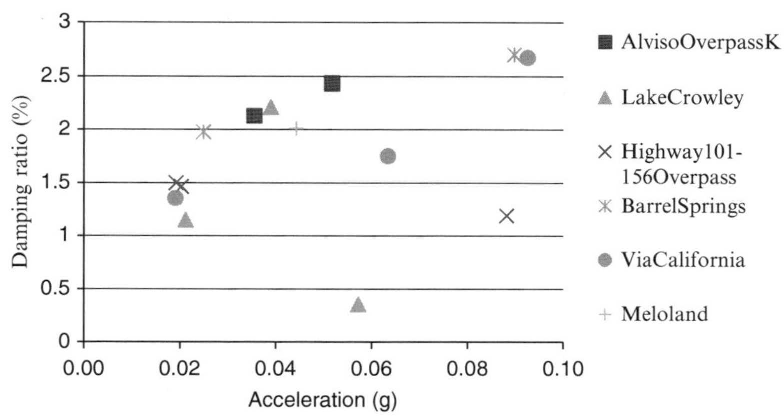

Fig. 11.7 Damping as function of max acceleration on deck estimated with EFDD

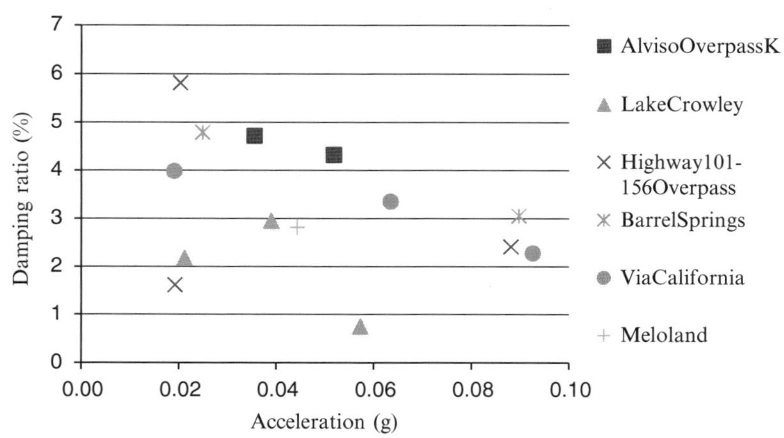

Fig. 11.8 Damping as function of max acceleration on deck estimated with SSI

up to 5.81 %. In general, although the upper bound estimated based on the SSI method is 6 %, there is not a clear trend in the results observed. The results obtained with the EFDD method seem to have a better agreement with the expectation of an increment as the acceleration increases.

11.5 Conclusions

The results observed in the assessment of the damping ratio using the simple 3-DOF's lumped mass model suggest an increment of the damping ratio with the amplitude of the external excitation. This observation is in good agreement to what was observed in the evaluated bridges, in which the damping ratio values – estimated by using the EFDD method– are below 2.5 % for low levels of vibration, produced by ambient vibrations or weak ground motion. However, according to the results obtained using ground motion data, this independency of the frequency content of the excitation is not directly extended to the lateral damping of bridges. These results show different damping ratios for ground motions with similar intensity applied to the same bridge. As far as the authors' knowledge, the bridges included in the database did not exhibit structural damage caused by the recorded events, even for the stronger ground motions. Therefore, further investigation is recommended to evaluate the inconsistency in the damping values estimated based on ground motion data calculated by the EFDD and the SSI methods.

Acknowledgements The first author acknowledges the support provided by the University of Medellin, Colombia. This study was also conducted with financial support from the Ministry of Transportation of British Columbia, Canada and the Natural Sciences and Engineering Research Council of Canada.

Appendix

Appendix 1 Analyzed bridges and their basic instrumentation information (Adapted from [5])

Station no.	Location	Bridge name	Bents angle	Structure/Characteristics	Instrumen-tation
1336	El Centro, Highway 8	Meloland Overpass	Perpen-dicular	2 spans. Concrete box girder connected monolithically to the abutments. Reinforced concrete central pier with one column. Timber piles as foundation system for a square concrete footing under central pier and also for abutments	Instrumented in 1978. Upgraded in 1991
13795	Capistrano Beach, I5	Via California	Perpen-dicular	6 spans. 6-cell concrete box girder. Cantilever abutments. Reinforced concrete columns, two columns per bent. Spread footings	Instrumented in 1999
24706	Palmdale, Highway 14	Barrel Springs	Skewed	5 spans. Concrete box girders. Open end diaphragm abutments. Two Rectangular concrete columns per bent. Spread footings support all bents and abutments. Bridge was retrofitted	Instrumented in 1994
47315	San Juan Bautista, Highway 101	156 Overpass	Skewed	6 spans. Welded steel girders with concrete diaphragms supported by rocker bearings. Seat abutments with catcher beams and integral wing walls. Concrete bents with two rectangular columns. Spread footings. Retrofitted by adding concrete diaphragm to girders and infill walls between the columns	First instrumented in 1977. Re-instrumented in 2002
54730	Lake Crowley	Highway 395 Bridge	Skewed	2 spans. Continuous concrete box girder. Diaphragm abutments. Bents with two circular concrete columns. Spread footing foundations	Instrumented in 1995
57748	Santa Clara, Highway 237	Alviso Overpass (Bridge K)	Skewed	6 spans. Continuous concrete box girder without intermediate hinges. Seat abutments with elastomeric bearings. One rectangular concrete column per bent with flares in the upper portion. Prestressed concrete piles	Instrumented in 1995
57748	Santa Clara, Highway 237	Alviso Overpass (Bridge L)	Skewed	6 spans. Continuous concrete box girder without intermediate hinges. Seat abutments with elastomeric bearings. Two rectangular concrete columns per bent with flares in the upper portion. Prestressed concrete piles	Instrumented in 1995
68717	Rohnert Park, Highway 101	Rohnert Park Bridge	Perpen-dicular	2 spans. Continuous concrete box girder. Sliding diaphragm abutments with elastomeric bearings and shear keys. Two semicircular concrete columns per bent, with flares in the upper portion. Concrete piles as foundation system for abutments bents	Instrumented in 1995
89324	Rio Dell, Highway 101	Painter St Overpass	Skewed	2 spans. Continuous concrete box girder. Diaphragm abutments. Two circular concrete columns per bent. Concrete piles	Instrumented in 1977
89708	Arcata, Highway 101	Murray Road Bridge	Perpen-dicular	4 spans. Six concrete T-beams supported on open end cantilever abutments and concrete bents. Abutments with rocker bearings. Two rectangular concrete columns per bent. Spread footings	Instrumented in 1995

Appendix 2 Analyzed ground motion records basic information (Adapted from [5])

No.	Station	Bridge	Epicenter location/Date	Epicentral dist (km)	PGA(g)	PGV (cm/s)	PGD (cm)
1	1336	Meloland Overpass	BorregoSprings 07Jul2010	120.2	0.012	1.05	0.137
2	1336	Meloland Overpass	Calexico 20Nov2008	50.4	0.017	0.61	0.158
3	1336	Meloland Overpass	Calexico 22May2010	35.2	0.031	0.67	0.068
4	1336	Meloland Overpass	Calexico 30Dec2009	41.2	0.174	16.61	3.899
5	1336	Meloland Overpass	Calexico 04Apr2010	58.9	0.213	19.05	13.928
6	13795	I5/Via California	Borrego Springs 17-07-2010	110.1	0.006	0.6	0.14
7	13795	I5/Via California	Calexico 04-04-2010	337.1	0.02	4	2.35
8	13795	I5/Via California	Chinohillls 29-07-2008	52.1	0.023	1.14	0.28
9	24706	Barrel Springs Bridge	BigBearCity 22022003	120.6	0.009	0.27	0.043
10	24706	Barrel Springs Bridge	Chinohills 29072008	73.6	0.027	0.97	0.063
11	47315	156 Overpass	SanJuanBautista 12012011	12.4	0.016	0.47	0.036
12	47315	156 Overpass	AlumRock 30102007	65.7	0.021	1.5	0.121
13	47315	156 Overpass	Aromas 02072007	4.5	0.073	2.9	0.198
14	54730	Hwy 395	QualeysCamp 18-09-2004	48.7	0.014	0.39	0.07
15	54730	Hwy 395	Toms Place 26-11-2006	16.1	0.015	0.25	0.01
16	54730	Hwy 395	Mammoth Lakes 12-06-2007	12.8	0.051	0.83	0.1
17	57748	Alviso Overpass (K)	Gilroy 13052002	60	0.077	0.66	0.074
18	57748	Alviso Overpass (K)	Gilroy 13052002	60	0.077	0.66	0.074
19	68717	Hwy 101 Bridge	Bolinas 1999	49	0.009	1.15	0.121
20	89324	Painter St Overpass	CapeMendocino 21111986	29.9	0.146	18.38	1.53
21	89324	Painter St Overpass	PetroliaAftershock2 26041992	40.8	0.198	33.44	8.419
22	89324	Painter St Overpass	PetroliaAftershock1 26041992	40.3	0.515	45.62	6.248
23	89324	Painter St Overpass	Petrolia 25041992	17.6	0.541	44.7	17.8
24	89708	Murray Road Bridge	Ferndale 09012010	64.7	0.077	13.12	3.149

References

1. Atkins JC, Wilson J (2000) Analysis of damping in earthquake response of cable-stayed bridges. In: Proceedings of the 12th world conference on earthquake engineering, Auckland, Paper no. 1468
2. Chopra AK (1995) Dynamics of structures. Prentice Hall, Upper Saddle River
3. Magalhaes F, Cunha A, Caetano E, Brincker R (2010) Damping estimation using free decays and ambient vibration tests. Mech Syst Signal Process 24:1274–1290
4. Dyke SJ, Spencer BF Jr, Quast P, Kaspari DC Jr, Sain MK (1996) Implementation of an AMD using acceleration feedback control. Microcomput Civ Eng Spec Issue Act Hybrid Struct Control 11:305–323
5. Fraino M, Ventura, CE, Liam Finn WD, Taiebat M (2012) Seismic soil-structure interaction effects in instrumented bridges. In: Proceedings of the 15th world conference on earthquake engineering, Lisbon
6. SVS, ARTeMIS Extractor Pro, Release 3.41, Structural Vibration Solutions, Aalborg, 1999–2004

Chapter 12
Vibration Monitoring and Load Distribution Characterization of I-89 Bridge 58-N

Geoff May and Eric M. Hernandez

Abstract The University of Vermont Structural Monitoring and Diagnostics Lab is collaborating with the Vermont Agency of Transportation to instrument and monitor traffic induced vibrations in bridge 58N on interstate 89 in the town of Richmond, VT. The bridge consists of an unorthodox concrete and steel deck supported on reinforced concrete piers. A critical portion of the deck is strategically instrumented with 24 strain sensors and 10 accelerometers. The main objective of the monitoring project is to use multi-scale vibration data to formulate a finite element model of the bridge deck that allows engineers to accurately assess the structural capacity of the deck under heavy traffic loads. Two methods are combined: (1) A direct method based on dynamic strain measurements at maximum stress locations in the deck and (2) An indirect method based on an updated sub-structured finite element model of the bridge deck updated based on frequencies and mode shapes identified from traffic induced acceleration measurements.

Keywords Bridges • Dynamics • Vibrations • Instrumentation

12.1 Motivation

Interstate 89 is the busiest highway in the state of Vermont, for the last 5 years an annual average of 3.5 million vehicles traveled through it [1]. The Vermont Agency of Transportation (VAoT) has expressed concerns regarding twin bridges 58N and 58S in Vermont's I-89. The main concern is the structural adequacy of the deck stringers to sustain heavy traffic loads. Material testing on the stringers shows a yielding strength that varies from 22 to 37 ksi, which is lower than the grade of steel typically required in bridges. In addition to the lower steel grade of the stringers, composite action between the steel elements (girders and stringers) and the reinforced concrete deck is uncertain since no information is available in the structural drawings regarding the compressive strength of concrete nor the connection elements that would enable shear stress transfer between the concrete and the steel beams.

Initial estimates based on standard codified procedures indicate a low rating for this bridge, however as it will be shown in the next section, the deck's structural system is unusual and the possibility exists that the codified procedures are not applicable to this unusual structural system. The VAoT is interested in gaining a better understanding of the static and dynamic behavior of this bridge in order to accurately estimate the load distribution pattern under traffic loads. VAoT has funded a project with the UVM Structural Monitoring and Diagnostics Lab to instrument, collect and analyze operational vibration data for a period of at least 1 year. The monitoring began March of 2012.

The main research question addressed in this paper is the inverse problem of extracting the information contained in the vibration measurements regarding the composite action taking place between the steel beams and the reinforced concrete deck. This will enable us to create an accurate finite element model of the deck which allows for prediction of structural response under heavy traffic loads.

G. May • E.M. Hernandez (✉)
College of Engineering and Mathematical Sciences, University of Vermont, Burlington 05405, VT, United States
e-mail: eric.hernandez@uvm.edu

A. Cunha (ed.), *Topics in Dynamics of Bridges, Volume 3: Proceedings of the 31st IMAC, A Conference on Structural Dynamics, 2013*, Conference Proceedings of the Society for Experimental Mechanics Series 38,
DOI 10.1007/978-1-4614-6519-5_12, © The Society for Experimental Mechanics, Inc. 2013

12.2 Review of Previous Work

In this section, we review some of the most recent and relevant work published by researchers in the area of vibration based structural health monitoring of steel beam-concrete slab bridges. We focus on two specific aspects: (1) identification of structural mechanisms of traffic load transfer (also referred to in the literature as load distribution factors) and (2) identification of dynamic response characteristics, namely, natural frequency, mode shapes and damping.

12.2.1 Estimation of Load Transfer Mechanisms from Measured Data

Kim and Nowak [2] used strain measurements induced by operational traffic loads to investigate conventional I-girder bridges load distribution factors and to assess AASHTO recommendations on dynamic impact factors [3]. The strains were measured on the bottom flange of I-girders. The study was conducted on two similar bridges during a span of 48 h and recorded data induced by approximately 900 trucks. The researchers concluded that the measured load distribution factors (LDF) and dynamic impact factors are consistently lower than the AASHTO recommendation. Breña et al. [4] monitored a conventional I-girder type highway overpass using strain measurements under a controlled live-load test. Strain measurements were used to estimate LDFs and these results were compared with the finite element model (FEM) load distribution factors. The researchers found that although the bridge did not have explicit shear transfer members (such as shear studs), the strain measurements across the cross section (assuming Bernoulli's hypothesis of linear strain distribution) was consistent with the condition of I-girders acting as composite with the reinforced concrete slab. Chakraborty and DeWolf [5] developed and implemented a continuous strain monitoring system on a three-span conventional composite I-girder overpass. The study reported on data over a period of 5 months. This study analyzed the LDFs using AASHTO recommendations, field measurements, and FEM analysis. The study also included the determination of the location of the height of the neutral axes when large trucks travel across the bridge. One of the conclusions of this study was that the measured strain levels and LDFs are typically significantly below those recommended by AASHTO. The authors stated that this is a byproduct of conservative simplifications typically used in conventional designs, such as not including redundancies, connection restraints, and the way in which loads are distributed to different parts of the structure. This conclusion is in agreement with previous studies [2]. Finally, Jauregui et al. [6] conducted controlled loading tests on a standard I-girder bridge and used strains measured at three points through the depth of the I-girders. Results of the investigation show that even though the girders did not have explicit elements to transfer shear forces in through the interface with the concrete slab, they behaved as partially composite sections in the positive moment region up to the onset of yielding.

In conclusion, the results presented in the existing literature typically use strain gauges to determine load distribution factors and the extent of composite action in typical slab-on-girder bridges. Strain data was measured either in controlled load tests or during operational traffic. In some cases LDFs from measurements were compared with FEMs and the AASHTO code.

12.2.2 Finite Element Model Updating

Finite element model (FEM) updating refers to the collection of computational methods that attempt to solve the inverse problem of estimating user-defined free parameters of a finite element model based on the minimization of a user-defined objective function of the residual between features extracted from measured vibration data on a system and model prediction of those features [7]. It is assumed throughout the process that the finite element model is consistent with the system, i.e., that there exists a set of free parameters that will minimize the residuals to a level consistent with the measurement noise (Morozov discrepancy principle). Numerous studies have been conducted in the realm of FEM updating for bridge evaluation and diagnosis. Here we will focus on discussing some of the most recent literature on FEM updating regarding slab-on-girder bridges.

Sanayei et al. [8] developed a baseline finite element model for bridge management using strain data collected during controlled static load tests. Feng et al. [9] applied a neural network-based system identification technique to establish and update a baseline FEM based on acceleration and strain measurements during operational traffic-induced vibrations. Brownjohn et al. [10] assessed the effects of a bridge strengthening. Experiment-based structural assessments consisting of long-term monitoring, a full-scale dynamic test, and FEM updating were conducted before and after the strengthening. Focusing on the dynamic testing and the modal analysis the effectiveness of the strengthening was evaluated through model updating. Hou and Lynch [11] employed a wireless monitoring system on a typical slab-on-girder bridge. Ambient vibration monitoring was conducted on the test bridge and the modal parameters were extracted. The modal assurance criterion was implemented to assess the correlation between modes found experimentally and those calculated by an updated FEM.

The vast majority of FEM updating involved in slab-on-girder bridge investigations exclusively used vibration data from accelerometers. Models were updated and calibrated by selecting various structural parameters to match the modal properties extracted from the vibration measurements. These studies typically employed dense sensor arrays across the entire length and width of the bridge.

In this project, we differ from previous studies in two main aspects: First, most of the studies have been carried out on standard I-girder bridges where each girder has similar stiffness. In our case, the bridge deck is atypical involving stiff longitudinal built-up girders which span the whole bridge length combined with more flexible and short longitudinal members (stringers) which span transverse floor beams supported on the stiffer longitudinal girders. Second, we explicitly combine acceleration and strain measurements to update the finite element model; in most of the previous studies (except for [9]), either accelerations or strains are exclusively used.

12.3 Bridge Description and Geometry

Bridges 58N & S are twin bridges and only bridge 58N was instrumented. The bridge was designed in 1961 and built in 1962. The bridge carries two lanes of traffic both in the north direction. As can be seen from Fig. 12.1, the deck is supported on two longitudinal built-up I-girders, which rest upon two intermediate reinforced concrete piers and at the ends on reinforced concrete abutments.

A cross section of the typical deck's structural system is depicted in Fig. 12.2. The reinforced concrete slab (7½″ thick) is supported at the ends on the main longitudinal girders with variable depth between 96″ (at the midspan) and 168″ (at the supports) and on continuous intermediate stringers (W18 × 60) parallel to the main girders. The stringers are supported on a transverse floor beam (W36 × 170) which itself is supported on the main girders. The transverse floor beams are spaced at 23 ft on-center.

Fig. 12.1 Overall view of bridge 58 N in I-89 Vermont

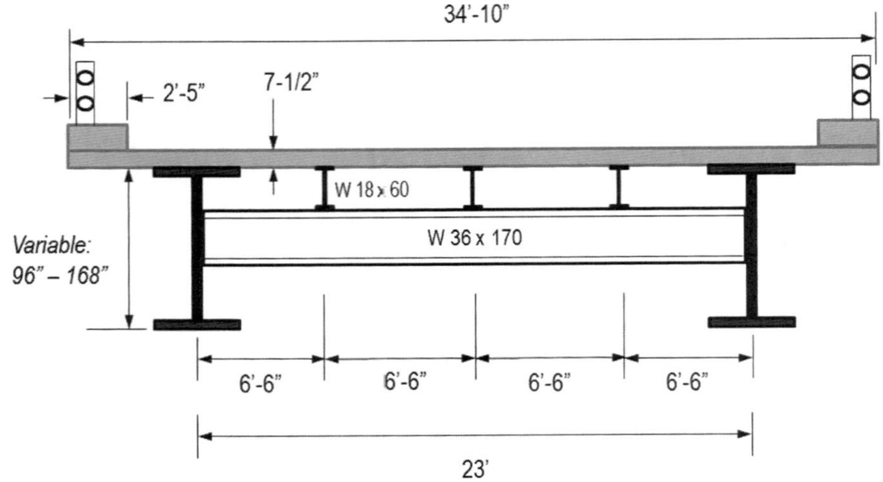

Fig. 12.2 Typical cross section of bridge deck with dimensions

12.4 Instrumentation

Instrumentation on the bridge consists of 10 accelerometers (PCB Model 393A03) and 24 strain sensors (PCB model 242A02) placed on the deck as shown in Fig. 12.3. The location of the sensors was selected to capture the maximum expected structural response of the stringers while at the same time minimizing the required number of sensors. It is worth noting that the main interest in the instrumentation is to capture the behavior of the stringers, however due to the unusual structural system of the bridge we will also capture the global structural response. This will be clearly illustrated in data section of the paper. To perform the data acquisition we used the LMS SCADAS MOBILE SCM05. A uniform sample rate of 200 Hz was selected for all sensors.

We began collecting data on the bridge on March 2012 and plan to extend data collection until the end of summer of 2013. Measurements consist of 1-h long records recorded sporadically various times a week and short records (< 1 min.) typically used for comparative purposes, these also include video. We have also performed some controlled tests under known moving loads at fixed traveling speeds of 40 and 60 mph.

12.5 Experimental Results

12.5.1 Strain Measurements

Twenty-four strain sensors are mounted at the top and bottom flange of the stringers at positions shown in Fig. 12.3. The method of mounting the strain sensors involves welding a 3 × 4 × 1 in. steel plate onto the flange and then mounting the strain sensor on the mounting plate. The mounting plate was necessary because the strain sensors require a tightening screw, which based on VAoT specifications, could not be drilled into the flanges, and thus an indirect strain measurement was necessary. Based on laboratory test results and detailed finite element modeling, we have found that the strain in the flange should be approximately 2.5 times the strain measured in the mounting plate.

As can be seen from Fig. 12.3, strain sensors are placed on the top and bottom flange and thus by invoking Bernoulli's linear strain assumption we can estimate the location of zero strain (neutral axis) by performing linear interpolation. Figure 12.4 shows the strain measurements in the top and bottom flange at position 7 due to the passing of 2 trucks of unknown axle weight and spacing. As can be seen, under varying degree of strain intensity, the neutral axis depth measured with respect to the top of the bottom flange is stable between 10 and 10.5 in.. Although beyond the scope of this paper, it was found that the neutral axis depth varies appreciably as a function of ambient temperature. Additionally, upon analysis of strain measurements, we have found that as we move from position 7 to position 1, the location of the identified neutral axis measured from the bottom flange of the stringers descends (see Fig. 12.3 for definition of coordinate system). We hypothesize that this is due to the added compression induced by the whole bridge section deforming in bending and thus compressing the concrete slab together with the stringers.

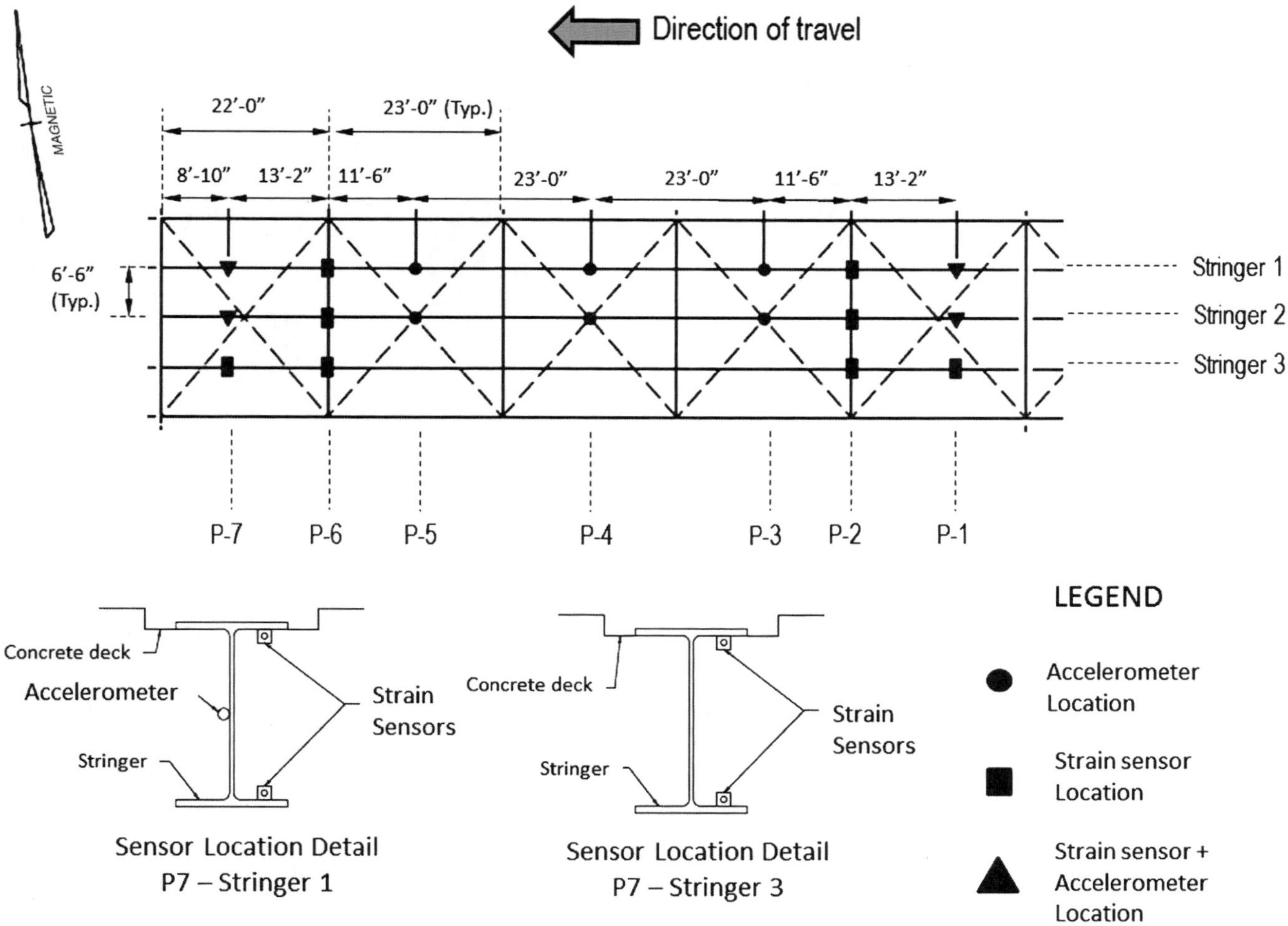

Fig. 12.3 Plan view of bridge deck instrumentation

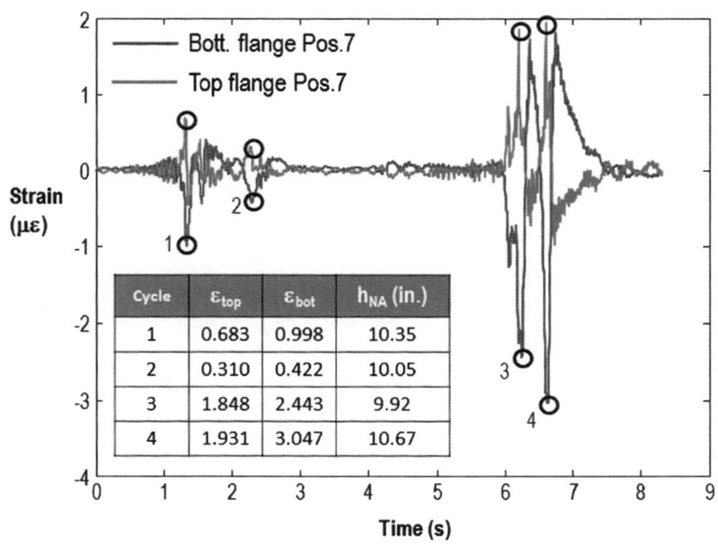

Cycle	ε_{top}	ε_{bot}	h_{NA} (in.)
1	0.683	0.998	10.35
2	0.310	0.422	10.05
3	1.848	2.443	9.92
4	1.931	3.047	10.67

Fig. 12.4 Unfiltered strain measurements at position 7 of stringer 2 due to multiple trucks of unknown weight (See Fig. 12.3 for coordinate system defining position of position 7)

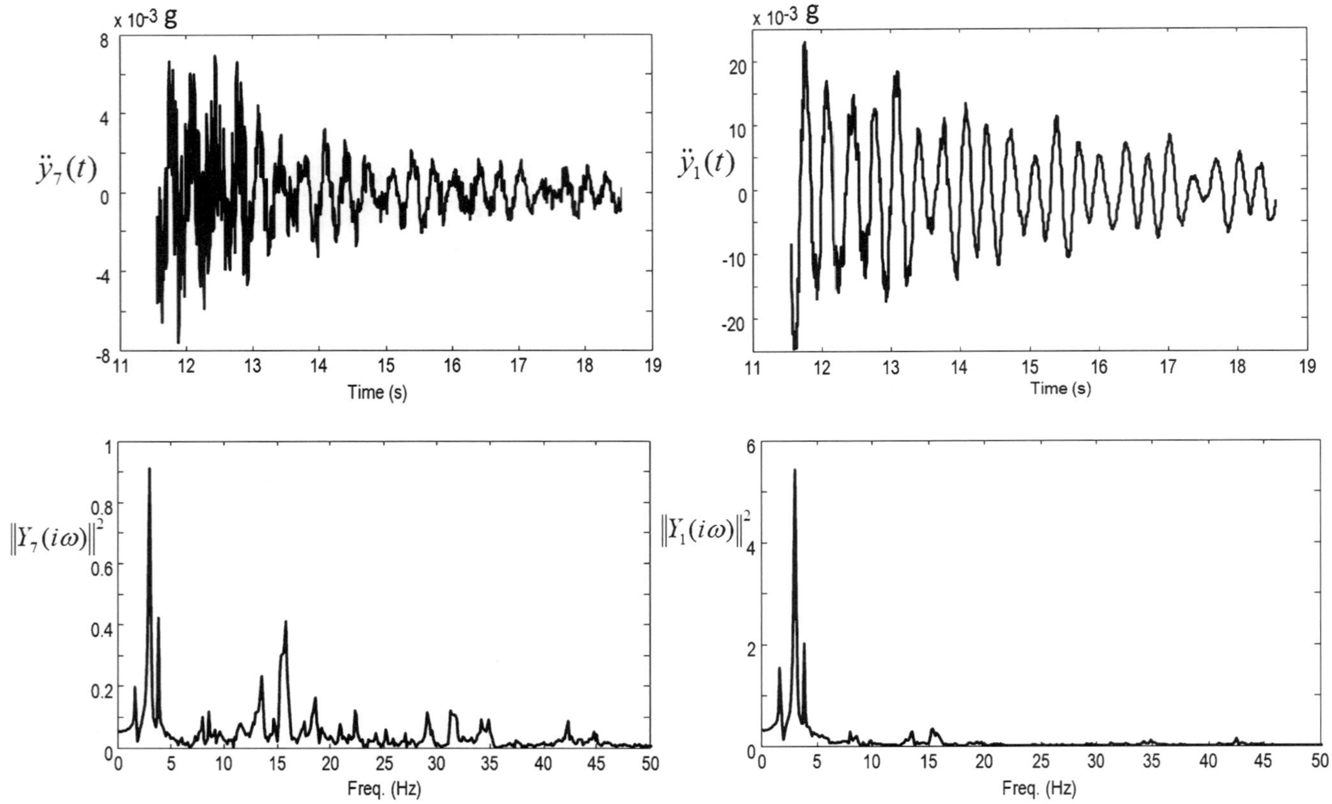

Fig. 12.5 Unfiltered acceleration measurements for stringer 2 in position 7 (on the *left*) and position 1 (on the *right*). On the *bottom*, we show the square of the magnitude of the Fourier transform for each measurement

12.5.2 Acceleration Measurements

Ten accelerometers are mounted on the web of the stringers as shown in Fig. 12.3. Figure 12.5 shows the results of acceleration measurements (in g's) at two different positions along stringer 2. Since in general we do not know the weight and axle spacing, we use only the acceleration data corresponding to free vibration (after the vehicle has exited the bridge). It is well known that for systems with small damping ($< 10\ \%$) the ordinate corresponding peaks of the magnitude of the Fourier transform of the free vibration response coincides very closely with the undamped modal frequency [12]. Figure 12.5 also shows the amplitude of the discrete Fourier transform for the acceleration measurement recorded at position 1 and 7 in stringer 2.

By comparing the spectra between position 1 and position 7 it was determined that vibration modes were occurring on a global scale as well as a local scale. Position 1 acceleration measurements were dominated by high amplitude, low frequency vibration modes. These modes correspond to the entire, global vibration of bridge attributed to the structural properties of the main exterior girders. At position 7 these same global modes were still recorded but at a much smaller amplitude. Position 7 acceleration measurements instead displayed high amplitude response in vibration modes at higher frequencies. These modes correspond to the local vibrations of the deck stringers. Table 12.1 illustrates the various identified modal frequencies.

12.6 Finite Element Modelling

Two finite element models were formulated and updated to investigate the structural behavior of the bridge. The first model uses frame elements to model the entire length without deck details. Thirty-eight frame elements were needed to sufficiently model the changing cross-section due to the parabolic haunch of the girders. Included in the girder cross-section was half the

Table 12.1 Identified modal frequencies versus updated FEM frequencies

Mode	Description	Modal frequency (Hz)	
		Identified	FEM
1	Vertical – girders	1.77	1.76
2	Vertical – girders	3.10	3.13
3	Vertical – girders	3.90	4.06
4	Vertical – stringers	13.68	13.69
5	Vertical – stringers	15.03	16.38
6	Vertical – stringers	18.44	17.74

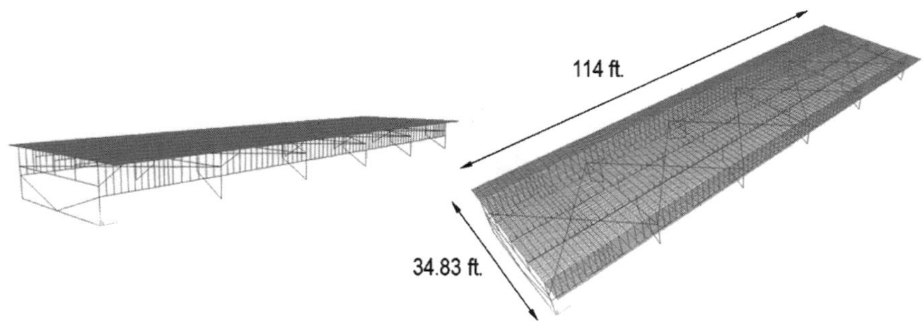

Fig. 12.6 Overview of three dimensional sub-structured finite element model with overall dimensions. The model is composed of frame elements representing girders, floor beams, stringers, rigid and flexible links as well as shell elements to model the reinforced concrete slab

width of the concrete road deck and additional line mass to approximate the additional dead weight associated with non-structural elements. Adding the concrete deck to the girder cross-section introduced composite action in the analysis.

The second model is a sub-structured finite element model of the structure that was generated using 1,141 frame elements and 10,341 shell elements. The model represents approximately 20 % of the length of bridge and contains the instrumented portion of the bridge. Figure 12.6 displays an overview of the detailed FEM. This model focuses primarily on the local vibration modes mostly engaging the stringers. Coupled stiffness matrices were placed at the free end of the model at the centroid of the exterior girder to include the stiffness of the remaining 80 % of the bridge (not being modeled explicitly). Horizontal springs were placed at the roller supports to model the elastic effect of the expansion joint at the abutment. Massless link elements connected the concrete deck shell elements to the frame elements of the model. Since the composite global girder model correlated well with the identified global modal frequencies the rigid links connecting the deck to the girders were taken as rigid. The link elements connecting the deck elements to the stringer frame elements were updated to better correlate with the identified local modal frequencies. The compressive strength of the concrete deck was kept constant at 4 ksi. Table 12.1 depicts the comparison between identified modal frequencies from the data and finite element models. The free parameters in the FEM to be updated were: (1) support springs (2) stiffness of link elements between stringers and slab (3) stiffness of link elements between girders and slab.

12.7 Conclusions

Based on the measured acceleration and strain data we can preliminarily conclude that even though the structural drawings do not specify any explicit member for shear transfer at the interface of girder and the concrete slab, vibration data suggests that the bridge main girders are acting completely composite with the concrete slab. One possible mechanistic explanation for this phenomenon might be the fact that bridge built-up girders possess rivets that protuberate on the top flange, thus creating a rugged interface which allows for significant shear transfer. In the case of the steel stringers, we have found that even though these do not possess any shear transfer members they are acting partially composite (at around 70%). These estimates allow engineers to estimate the member capacity under normal traffic loads and to compute upper bounds on the member ultimate bending capacity. Table 12.2 presents preliminary relative LDF values for the three stringers under two recorded load tests and three FEMs results considering various loading positions in the transverse direction. We have also found as other researchers in the past have, that AASHTO load distribution factors for the stringers are overly conservative, in our case by a factor of approximately 40%.

Table 12.2 Comparison of relative load distribution factors

	Relative load distribution factor		
	Stringer-1	Stringer-2	Stringer-3
FEM – case 1	1.00	0.92	0.24
FEM – case 2	0.90	1.00	0.35
FEM – case 3	1.00	0.75	0.17
Load test – 1	1.00	0.81	0.02
Load test – 2	1.00	0.89	0.03

Acknowledgements The research project presented in this paper was funded by the Vermont Agency of Transportation (VAoT). Special thanks to Jason Tremblay of VAoT for his support and coordination efforts during the instrumentation process. Conclusions contained herein are those of the authors and do not necessarily represent the views of the Vermont Agency of Transportation.

References

1. Web-based traffic data visualization http://www.wimweb.net
2. Kim S, Nowak AS (1997) Load distribution and impact factors for I-girders bridges. ASCE J Bridge Eng 2(3):97–104
3. American Association of State Highway and Transportation Officials (AASHTO) (1994) LRFD bridge design specifications. American Association of State Highway and Transportation Officials (AASHTO), Washington, DC
4. Breña SF, Jeffrey AE, Civjan SA (2011) Evaluation of a non-composite steel girder bridge through live-load field testing, ASCE Journal of Bridge Engineering, doi: 10.1061/(ASCE)BE.1943-5592.0000398
5. Chakraborty S, DeWolf J (2006) Development and implementation of a continuous strain monitoring system on a multi-girder composite steel bridge. ASCE J Bridge Eng 11(6):753–762
6. Jauregui DV, Yura JA, Frank KH, Wood SL (2002) Field evaluation of decommissioned noncomposite steel girder bridge. ASCE J Bridge Eng 7(1):39–49
7. Friswell M, Mottershead JE (1995) Finite element model updating in structural dynamics. Kluwer, Dordrecht, Netherlands
8. Sanayei M, Phelps JE, Sipple JD, Bell ES, Brenner BR (2012) Instrumentation, nondestructive testing and finite element model updating for bridge evaluation using strain measurements. ASCE J Bridge Eng 17(1):130–138
9. Feng MQ, Kim DK, Yi J-H, Chen Y (2004) Baseline models for bridge performance monitoring. ASCE J Eng Mech 130(5):562–569
10. Brownjohn JMW, Moyo P, Omenzatter P, Lu Y (2003) Assessment of highway bridge upgrading by dynamic testing and finite element model updating. ASCE J Bridge Eng 8(3):162–172
11. Hou T, Lynch JP (2006) Rapid-to-deploy wireless monitoring systems for static and dynamic load testing of bridges: validation on the grove street bridge. In: SPIE 13th annual international symposium on smart structures and materials, San Diego
12. Tedesco JW, McDougal WG, Ross CA (1999) Structural dynamics: theory and applications. Addison Wesley, Menlo Park

Chapter 13
Dynamic Modeling and Field Testing of Railroad Bridges

Hani Nassif, Peng Lou, and Ying-Jie Wang

Abstract In this paper, results for a study to investigate the impact of increasing the weight of freight railcar from 1,170 kN (263 kips) to 1,272 kN (286 kips) on typical bridges that are part of the New Jersey rail network are presented. Based on the field inspection reports, a number of critical bridges on New Jersey's rail lines were selected and load-rated based on the current American Railway Engineering and Maintenance-of-Way Association (AREMA) specifications. Two-Dimensional (2D) dynamic models and field instrumentation and testing were adopted for the more accurate assessment of these bridges and to develop a refined methodology for evaluating and load-rating railroad bridges. The field study included instrumentation and testing under moving freight and passenger railcars. The steel bridge is simulated as a Bernoulli-Euler beam and the moving train is modeled using rigid-body dynamics. The method of modal superposition is adopted to compute the dynamic effects of the train-bridge interaction system. The dynamic model was validated with results from the field tests. The impact factor for fatigue of the bridge under moving freight and passenger train were compared with AREMA Mean Impact Factor. Results show that the impact factor for bridge fatigue is increased by the free vibration component. Moreover, for passenger car, when the running speed is above 170 km/h, the impact factor for fatigue is slightly larger than the AREMA Mean Impact Factor.

Keywords Railroad bridge • Filed testing • Dynamic model • Impact factor • Fatigue

13.1 Introduction

By the late 1990s, economic and population growth resulted in significant freight movement. It is expected that the rail freight traffic will grow sharply for the next 20 years. Therefore substantial demand will be put on the already heavily-used railroad system. The freight railroad system enables cost-effective movement of large volume of freight, and is important because the alternative transportation methods, vehicles and trucks, cause concerns about congestion, air quality, and safety. However, the cost to build and maintain infrastructure and equipment is very high, and it is very difficult to make long-term investment in railroad infrastructure. Additionally, many railroad bridges were built before World War II and are approaching their design life limit, which creates additional concerns. In New Jersey freight railcars utilize portions of passenger rail system to reach their destinations, sharing lines with NJ Transit commuter rail service.

Fatigue phenomenon influences service life and maintenance costs of steel railroad bridges, which is caused by accumulation of live load stress over a long period. During bridge design and evaluation, the dynamic effect of bridges resulting from moving train is usually expressed as impact factor. However, the impact factor defined for fatigue differs from various codes and specifications. The AASHTO LRFD [1] recommends 15% of the live load for the fatigue limit state. Some

H. Nassif, Ph.D., P.E. (✉)
Rutgers Infrastructure Monitoring and Evaluation (RIME) Laboratory Rutgers, The State University of New Jersey
96 Frelinghuysen Road, Piscataway NJ 08854, United States

Department of Civil Engineering, College of Engineering, Kyung Hee University
e-mail: nassif@rci.rutgers.edu

P. Lou, Ph.D. Candidate • Y.-J. Wang, Ph.D.
Rutgers Infrastructure Monitoring and Evaluation (RIME) Laboratory Rutgers, The State University of New Jersey
96 Frelinghuysen Road, Piscataway NJ 08854, United States

A. Cunha (ed.), *Topics in Dynamics of Bridges, Volume 3: Proceedings of the 31st IMAC, A Conference on Structural Dynamics, 2013*, Conference Proceedings of the Society for Experimental Mechanics Series 38,
DOI 10.1007/978-1-4614-6519-5_13, © The Society for Experimental Mechanics, Inc. 2013

Table 13.1 Assumed mean impact load percentages (from AREMA 2009 Table 15-1-8)

Member	Percentage (%)
Member with loaded lengths equal or less than 10 ft (3 m) and no load sharing	65
Hangers	40
Other truss members	65
Beams, Stringers, Girders and Floor beams	35

Note: Where bridges are designed for operation of trains handling a large percentage of cars with flat or out of round wheels which increase impact and/or poor track which increases impact, and the loaded length of the member is less than 80 ft (24 m), the mean impact should be 100 % of the design impact

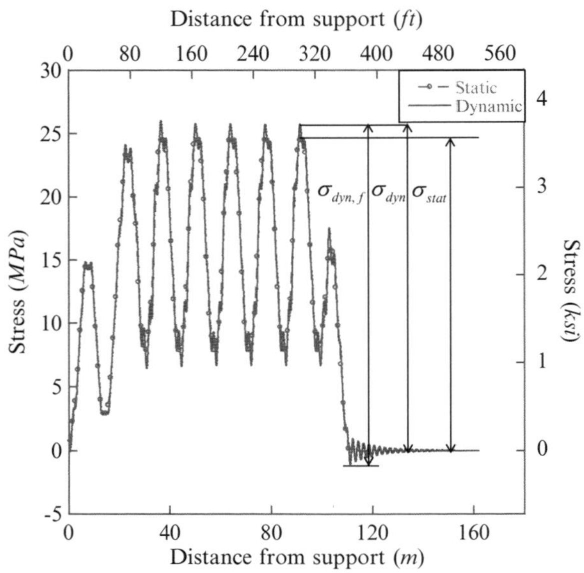

Fig. 13.1 Example of stress time history

codes define the impact factor for fatigue as percentages of the static values of the dynamic impact factor. For example, the Korean Railway Bridge Design Code [2] specifies the fatigue factor as 65% for bridges longer than 9 m, and the Japanese Railway Standards for Steel Bridge [3] specifies 75%. Recently, Lee et al. [4] suggested that the impact factor for fatigue can be only half of that used in ordinary static design of the present Korean code. The AREMA Specifications 2009 [5] defined the impact factor for fatigue as the "Mean Impact Load", calculated as a percentage of the dynamic impact factor used in static design. AREMA has specified different values for different bridge members as shown in Table 13.1.

The major factors governing fatigue of steel bridges are the numbers of stress cycles, the magnitude of the stress range, and the relevant Fatigue Detail Category. Therefore, in fatigue design or evaluation, the maximum stress range rather than the maximum stress is considered. Figure 13.1 shows typical dynamic and static stresses in a bridge member due to the passage of a train passing over the bridge. For the normal static design, the impact factor (IM) is normally calculated using (13.1). However, for steel railroad bridges, the impact factor for fatigue is defined as (13.2).

$$IM = \frac{\sigma_{dyn}\text{-}\sigma_{stat}}{\sigma_{stat}} \tag{13.1}$$

$$IM = \frac{\sigma_{dyn,f}\text{-}\sigma_{stat}}{\sigma_{stat}} \tag{13.2}$$

where, σ_{stat} is the maximum static stress, σ_{dyn} is the maximum dynamic stress, $\sigma_{dyn,f}$ is the maximum dynamic stress range.

In this paper, typical rail bridges on various New Jersey rail lines were reviewed to investigate the impact of the increased railcar weight. A two-dimensional (2D) train-bridge dynamic model is developed and validated using results from field tests. Using the validated model, the impact factor for fatigue of a typical steel plate girder bridge for different speeds of the train was determined.

13.2 Field-Testing Methods

In this study, various field testing systems were utilized to obtain the actual strain, deflection and velocity data. Various types of sensors and data acquisition systems were instrumented on a selected 12.2 m long, open deck steel girder railway bridge. The bridge supports two active tracks using four girders that allow two-way traffic. The field instrumentation includes Structural Testing System (STS) and the Laser Doppler Vibrometer (LDV). A typical testing plan is shown in Fig. 13.2.

The Structural Testing System (STS), which measures the strain in structural members, is a modular data acquisition system manufactured by Bridge Diagnostics, Inc. (BDI) of Boulder, Colorado. The system includes three parts: strain transducers, junction nodes, and a main unit. The strain transducers shown in Fig. 13.3a were mounted on structural elements and connected with the junction nodes. Each strain transducer is assigned a unique number that can help to recognize the location of the transducer easily in the computer software when operating the test. The junction nodes shown in Fig. 13.3b can transfer the data from strain transducer to the main unit (Fig. 13.3b) that produces wireless signal. The junction nodes were placed on the pier of the tested span. The main unit is a processing unit that samples data and was placed between the junction nodes and the computer to allow control the system remotely by the operator.

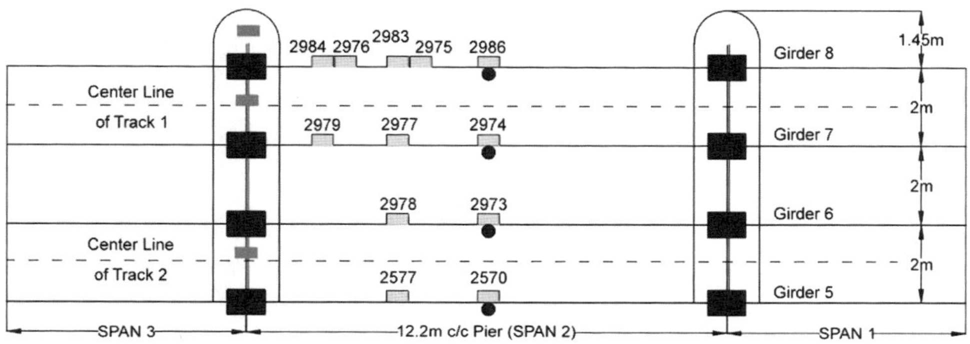

Fig. 13.2 Field testing plan for a typical span of selected bridge

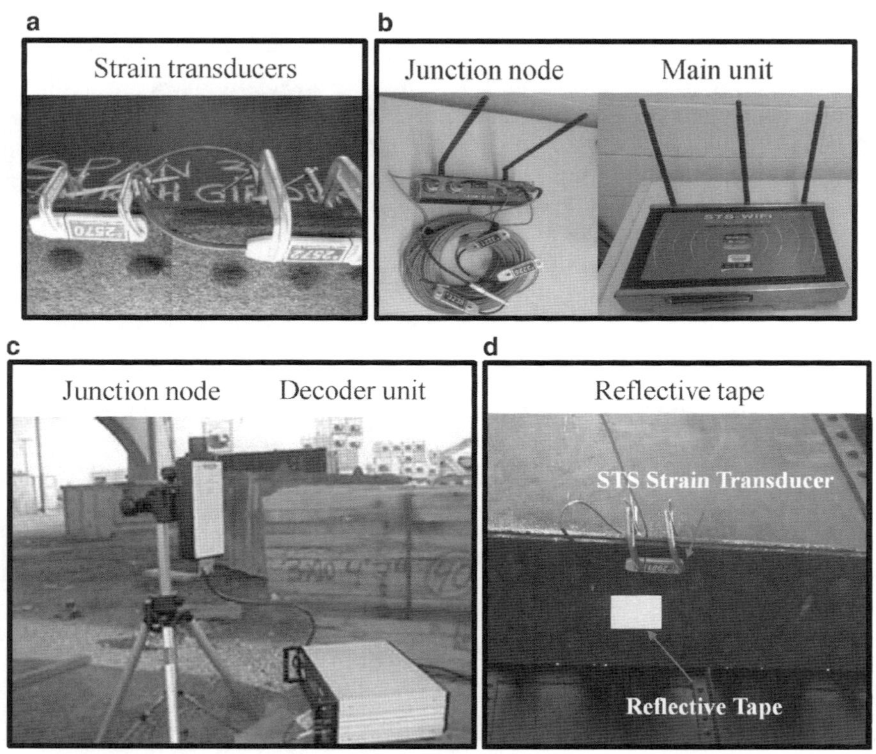

Fig. 13.3 (**a**) Strain transducers on the girder; (**b**) Junction nodes and main unit of STS; (**c**) Laser Doppler Vibrometer mounted to a tripod, Decoder unit; (**d**) Reflective tape on structural element

The Laser Doppler Vibrometer (LDV) is a non-contact measuring device that measures displacement and velocity of a remote point in structural members. The system is composed of three parts: the helium neon Class II laser head (Fig. 13.3c), the decoder unit (Fig. 13.3c), and the reflective target (Fig. 13.3d) attached to the structure. The laser head is mounted to a tripod that is positioned underneath the target.

13.3 2-D Dimentional Dynamic Model and Model Validation

Figure 13.4 shows a two-dimensional (2D) model which was developed to simulate the dynamic behavior of the bridge-train interaction The moving train composed of several vehicles, including locomotive car and freight cars or passenger cars, runs over a simple supported bridge at the constant speed v. It is assumed that the wheel sets of each vehicle are kept in full contact with the bridge at all times. This assumption is made to ensure that the dynamic responses of the bridge and vehicle are linearly coupled, which can be computed using conventional time integration methods without iterations.

As shown in Fig. 13.4, each vehicle is composed of one car body, two identical bogies, and four identical wheel sets. The wheel sets and bogies are connected by primary suspensions, and the bogies and the carbody are connected by secondary suspensions, which are modeled as a linear spring-dashpot units [6]. The car body and two bogies are each assigned two DOFs, which are vertical displacement and rotation about the center point. Using rigid-body dynamics method, the equations of motions for vehicle components can be given as a series of second-order ordinary differential equations in the time domain.

In this study, the simple bridge is modeled as a linear elastic Bernoulli-Euler beam with identical sections. Using modal superposition method, the equation of motion for the bridge subjected to moving train can be written as a series of second-order ordinary differential equations with generalized displacements.

By combining vehicle equations and bridge equations together, the equations of motion for train-bridge system can be presented in a matrix form as

$$[M]\{\ddot{U}\} + [C]\{\dot{U}\} + [K]\{U\} = \{F\} \tag{13.3}$$

where [M], [C], [K] denote the mass, damping and stiffness matrices; $(\{U\}, \{\dot{U}\}, \{\ddot{U}\})$ are the vectors of displacement, velocity, and acceleration, respectively; and {F} represents the vector of exciting forces applied to the dynamic system.

To compute both the dynamic responses of the simple bridge and moving vehicles, the generalized matrix equation of motion given in (13.3) will be solved using Newmark-β method. In this study, $\beta = ¼$ and $\gamma = ½$ are selected, which implies a constant acceleration with unconditional numerical stability.

To validate the 2-D dynamic model described earlier, the field testing results including deflection, velocity, and strain data are compared with the dynamic model results. As shown in Fig. 13.5, the deflection and velocity results for the selected bridge model show good agreement (within 10 %) with the field testing data.

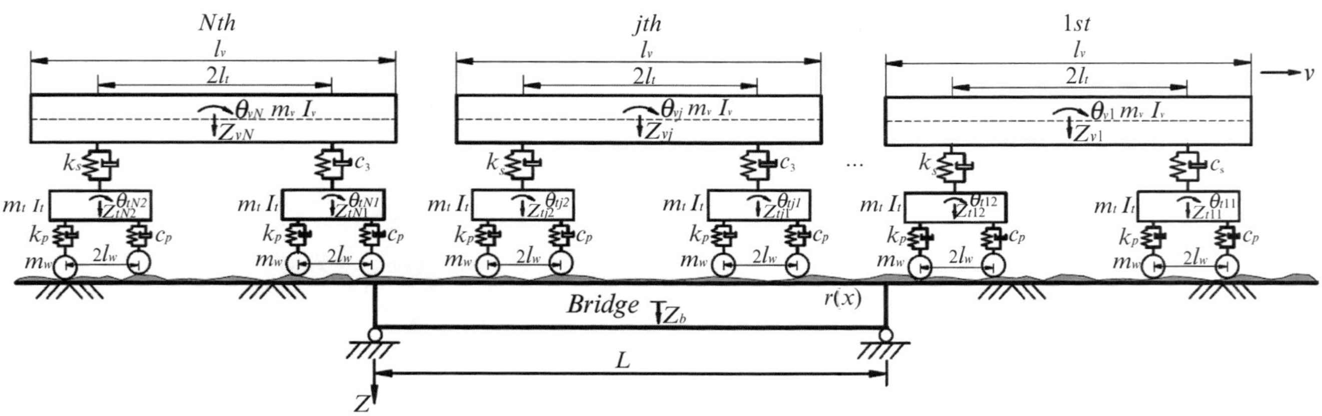

Fig. 13.4 2D train-bridge system model

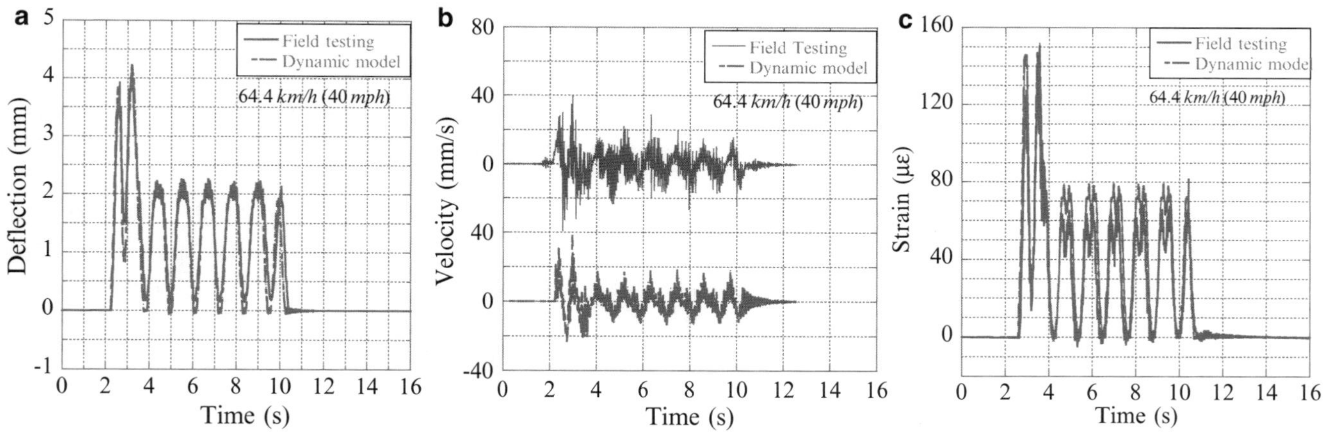

Fig. 13.5 Model validation (**a**) Deflection data, (**b**) Velocity data, (**c**) Strain data

13.4 Comparison with Arema Specifications

The dynamic effects of moving vehicles on bridges are important in the design and evaluation of their overall performance. In AREMA [5], for steel railway bridge, (13.4) was proposed to account for the vertical effect of the dynamic effect excluding the rocking effect (20% of live load).

$$IF = \begin{cases} 40 - \dfrac{3L^2(ft)}{1600} = 40 - \dfrac{3L^2(m)}{148.6} \Rightarrow L \leq 80ft = 24m \\ 16 + \dfrac{600}{L(ft) - 30} = 16 + \dfrac{182.9}{L(m) - 9.1} \Rightarrow L > 80ft = 24m \end{cases} \tag{13.4}$$

where L is the span length.

For train speeds below 96.6 km/h (60 mph), for all spans carrying equipment without hammer below, and for all spans other than truss spans carrying equipment with hammer below, the values of the vertical effects of the impact equations shall be multiplied by the following factor:

$$1 - \frac{0.8}{2500}\left(60 - \frac{62}{100}S(km/h)\right)^2 = 1 - \frac{0.8}{2500}(60 - S(mph))^2 \geq 0.2 \tag{13.5}$$

where S is the running speed.

Figure 13.6a show the impact factor of the selected bridge under the effect of a passing 1,272 kN (286 kips) freight car. For the static design, the maximum impact factor is 5% while the maximum impact factor for fatigue is 10%. Figure 13.6b shows that the maximum values for dynamic impact factor and fatigue impact factor are 11% and 22%, respectively. This suggests that the value for the fatigue impact factor is twice as high as the dynamic impact factor. This can be attributed to the free vibration component. When the freight car runs over the bridge, both the impact factor for fatigue and dynamic impact factor are below the value from AREMA Specifications. However, for the passenger car, when the running speed is above 170 km/h, the impact factor for fatigue is slightly larger than the AREMA Mean Impact Factor. This suggests that AREMA would slightly underestimates the dynamic amplification in the fatigue design at higher speeds. Therefore, for passenger car, the speed restriction should be applied to reduce the stress range associated with dynamic amplification.

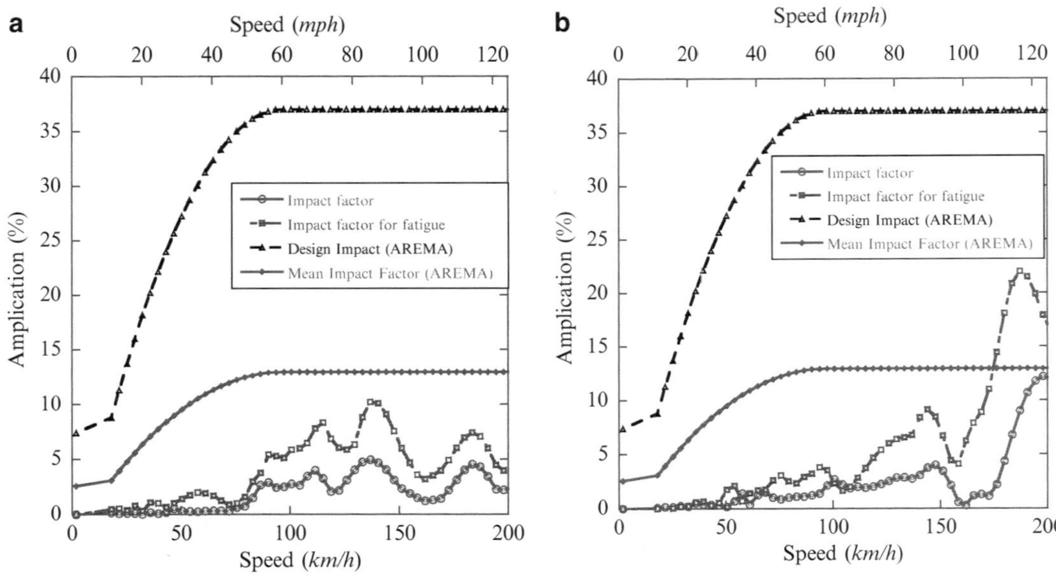

Fig. 13.6 Impact factor (**a**) Freight car, (**b**) Passenger car

13.5 Conclusions

In this study, typical rail bridges were used to investigate the dynamic impact of the increased freight railcar weight. Various types of equipment were used such as the Structural Testing System (STS), Laser Doppler Vibrometer (LDV), to monitor the dynamic responses (strain, deflection, and velocity) of the bridges under moving trains. A two-dimensional (2D) dynamic model of the train-bridge dynamic interaction system was developed and validated with field testing results. The impact factor for fatigue of the bridge under moving freight and passenger train were compared with AREMA Mean Impact Factor, the following conclusions can be summarized:

1. The impact factor for fatigue of the bridge is increased by the free vibration component.
2. For passenger car, when the running speed is above 170 km/h, the impact factor for fatigue is slightly larger than the AREMA Mean Impact Factor.

References

1. (2010) AASHTO LRFD bridge design specifications
2. Korea Rail Network Authority (2004) Railway bridge design code
3. Railway Technical Research Institute (2000) Japanese railway standards for steel bridges. Maruzen, Tokyo
4. Lee HH, Jeon JC, Kyung KS (2012) Determination of a reasonable impact factor for fatigue investigation of simple steel plate girder railway bridges. Eng Struct 36:316–324
5. AREMA (2006) Manual of engineering. American Railway Engineering and Maintenance-of-Way Association, Lanham
6. Yang YB, Yau JD, Wu YS (2004) Vehicle-bridge interaction dynamics: with applications to high-speed railways. World Scientific, Singapore

Chapter 14
Optimal Sensor Placement for Structures Under Parametric Uncertainty

Rafael Castro-Triguero, Senthil Murugan, Michael I. Friswell, and Rafael Gallego

Abstract This paper examines the influence of parametric uncertainties on the optimal sensor placement methodologies for modal analysis of a truss bridge. Four classical sensor location methodologies are employed: two based on the Fisher information matrix and two based on energy matrix rank optimization. Young's modulus, mass density and cross sectional dimensions are considered as uncertain parameters. The independent effects and cumulative effects of these uncertain variables on the sensor configuration are studied. The optimal locations of sensors under parametric uncertainty are assessed by the use of three different criteria. Furthermore, the robustness of this configuration is investigated for different levels of signal-to-noise ratio. The numerical results show the parametric uncertainties have significant influence on the optimal sensor configuration of a truss bridge.

Keywords Sensor placement • Uncertainty • Experimental/operational modal analysis

14.1 Introduction

Sensor placement in a host structure is an important initial step in the field of experimental modal analysis (EMA) and in particular in operational modal analysis (OMA) [1, 2]. Applications with wired and wireless methodologies necessarily need a planned methodology to locate the sensors. This sensor location can either be permanent for permanent monitoring, or can be temporary in the case of roving sensor monitoring. Furthermore, Structural Health Monitoring (SHM) techniques require an optimal number of sensors for system identification, structural damage identification, and finite element model updating.

Different methodologies have been developed for optimal sensor placement (OSP) [3]. These can be primarily classified into two groups: sub-optimal methods and methods based on formal optimization strategies. The sub-optimal methods are based on iterative techniques and the latter treat the problem as a classical optimization problem. Kammer [4] proposed an iterative method, the Effective Independence method (EFI), based on the maximization of the determinant of the Fisher Information matrix (FIM). The principal idea of this method is based on the linear independence of the mode shapes. The number of sensors is reduced in an iterative way by deleting the degrees-of-freedom (DOFs) from the mode shape matrix which contribute less to this independence. This method produces a sub-optimal solution. The Fisher matrix can also be weighted by the use of the mass matrix from a finite element model [5]. Energy matrix rank optimization (EMRO) techniques are similar to the EFI method but in this case the main idea is to maximize the strain energy of the structure rather than the determinant of the FIM matrix [6]. This method can also be based on the kinetic energy by using the mass matrix instead of the stiffness matrix [7].

In recent years, much research is focused on incorporating the uncertainties associated with real structures in the numerical simulation for reliable predictions. Uncertainties can be mainly classified as parametric uncertainty and model

R. Castro-Triguero (✉)
University of Cordoba, Cordoba CP 14071, Spain
e-mail: me1catrr@uco.es

S. Murugan • M.I. Friswell
College of Engineering, Swansea University, Swansea SA2 8PP, UK

R. Gallego
University of Granada, Campus de Fuentenueva, Granada CP 14081, Spain

A. Cunha (ed.), *Topics in Dynamics of Bridges, Volume 3: Proceedings of the 31st IMAC, A Conference on Structural Dynamics, 2013*, Conference Proceedings of the Society for Experimental Mechanics Series 38,
DOI 10.1007/978-1-4614-6519-5_14, © The Society for Experimental Mechanics, Inc. 2013

form uncertainty [8, 9]. The parametric uncertainties are associated with the discrepancies between the values of actual physical systems and the input parameters used for the analysis. Model form uncertainties are associated with the lack of accurate modeling of the physical system. However, in this paper, only the parametric uncertainties are considered. Uncertainty analysis makes use of statistical sampling, hypothesis testing and input–output data analysis to characterize the effect of parametric and model form uncertainties on the structural experiments or numerical simulation [8, 10–13]. The uncertainties can alter the optimal locations of sensors and a posteriori affect the structural health monitoring. However very few studies have considered the optimal placement of sensors with parametric model uncertainties.

In the present article the influence of parametric uncertainties on the optimal sensor placement methodologies for the modal analysis of a truss bridge is investigated. Four classical sensor location methodologies are employed: two based on the Fisher information matrix and two based on energy matrix rank optimization. Young's modulus, mass density and cross sectional dimensions of each member of the truss bridge are considered as uncertain parameters. The independent and cumulative effects of these uncertain variables on the sensor configuration are studied. The optimal locations of sensors under parametric uncertainty are assessed by the use of three different criteria: the Fisher matrix determinant, the modal assurance criterion error and the singular value decomposition ratio. Furthermore, the robustness of this configuration is investigated for different levels of signal-to-noise ratio.

14.2 Sensor Placement Algorithms

Four optimal sensor methodologies which are commonly used in OSP are considered in this study and discussed in this section. The two are based on the Fisher information matrix (EFI and EFIWM) and the other two based on the concept of energy matrix rank optimization (KEMRO and SEMRO).

14.2.1 Effective Independence Method (EFI)

Kammer [4] introduced the EFI optimal sensor placement algorithm which aims to search the best set of DOFs locations from all the candidate locations in the structure such that the linear independence of the mode shapes is maintained. The starting point of this method is the full modal matrix (Φ) from a finite element model. All the DOFs used in the FE model cannot be measured in the real structure due to physical limitations. Therefore, the DOFs corresponding to rotations and coordinates which cannot be measured are eliminated from the full modal matrix. Similarly, not all of the mode shapes can be experimentally measured, hence some target modes are selected to be optimally detected. Hence, the rows corresponding to DOFs that can be measured are kept and the columns corresponding to target modes are retained in the full modal matrix. The Fisher information matrix (FIM) is defined as

$$FIM = \Phi^T \ \Phi \tag{14.1}$$

If the determinant of the FIM is zero, the columns of the modal matrix (i.e. the target modes) are linearly dependent. Therefore, the purpose of the EFI method is to select the best DOFs (to place the sensors) which maximizes the determinant of the FIM.

A mass weighting version of the EFI (EFIWM) method is also studied in the literature [2]. In this case the FIM matrix corresponds to

$$FIM = \Phi^T \ M \ \Phi \tag{14.2}$$

14.2.2 Energy Matrix Rank Optimization (EMRO)

The basic idea underlying the EMRO algorithm is to achieve a sensor location configuration that maximizes the strain energy (SE) of the measured mode shapes from the structure [6]. The strain energy is

$$SE = \Phi^T \ K \ \Phi \tag{14.3}$$

where K is the stiffness matrix from the finite element model. This method is a weighted version of the EFI method where the FIM is assembled using a Cholesky decomposition of the stiffness matrix given by

$$K = C^T \, C \tag{14.4}$$

where C is a upper triangular matrix. In this case, the FIM is defined as

$$FIM_{SE} = \Psi^T \, \Psi \text{ where } \Psi = C \, \Phi \tag{14.5}$$

The procedure is the same as the EFI algorithm but Guyan reduction is employed to reduce the stiffness matrix at each step. This method is usually called SEMRO. Similar to the methodology based on the strain energy, the kinetic energy (KE) could be employed. This method is called KEMRO.

14.3 Numerical Results

The effects of uncertainty on the OSP methodologies are studied with a truss bridge structure shown in Fig. 14.1. A simple two-dimensional finite element model was used to simulate the dynamic behaviour of the truss bridge. The finite element analysis was carried out with the finite element code, CALFEM [14]. The bridge is discretized into 31 classical bar finite elements with 2-DOF/nodes which results in a total of 14 nodes and 28 DOFs. The first ten mode shapes of the 2D-truss bridge structure are considered as the target modes and the numerical analysis is performed with these modes.

14.3.1 Monte Carlo Simulation (MCS)

Uncertainty analysis can be performed with several stochastic methods available in the literature [8]. Monte Carlo simulation (MCS) is the most accurate and popular uncertainty analysis technique [9, 11, 12]. The main advantage of MCS is that uncertainty analysis can be performed without any modification of the deterministic finite element analysis program. The MCS requires a large number of simulations and results in a high computational cost. However, the computational cost or time for the truss type structure considered in this study is very small and therefore MCS is used in this study.

The uncertainty analysis of truss bridge with MCS is discussed in the following paragraphs.

1. *Selection of the parameters under uncertainty*. In the real world, the truss bridge structure has several parameters which are uncertain in nature. The parameters related to geometry and material properties of the bars can be uncertain because of the error tolerances and thermal treatments during their manufacture, or variations in the ambient temperature. In the present study, material properties and the cross sectional dimensions of the bars are considered as random variables with Gaussian distributions. The mean values and coefficients of variation (CoV) of these uncertain parameters are selected from Ref. [14] and given in Table 14.1.
2. *Generation of samples for the stochastic parameters*. A simple random sampling scheme using MATLAB is used to generate the random samples of the uncertain parameters. The Young's modulus (E), mass density (ρ) and cross section area (A) of the bars are considered as independent random variables and the random samples are generated. Monte Carlo

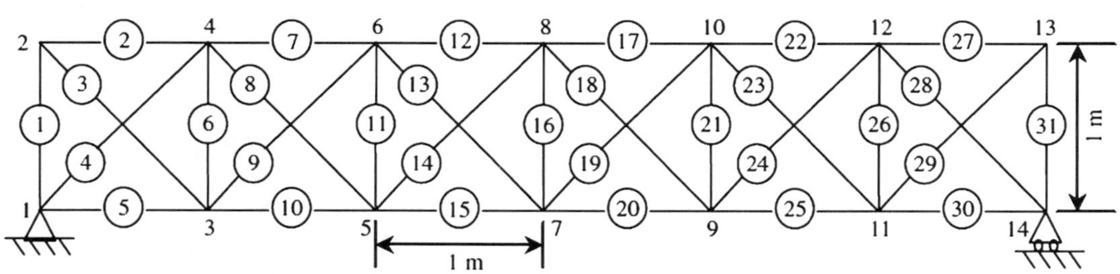

Fig. 14.1 Two dimensional truss bridge

Table 14.1 Stochastic properties of 2D-truss bridge parameters

Stochastic properties	Mean	C.o.V. (σ/μ) (%)	Distribution
Young's modulus, GPa	70	8	Normal
Mass density, kg/m^3	2,800	4	Normal
Cross sectional area, m^2	0.0025	5	Normal

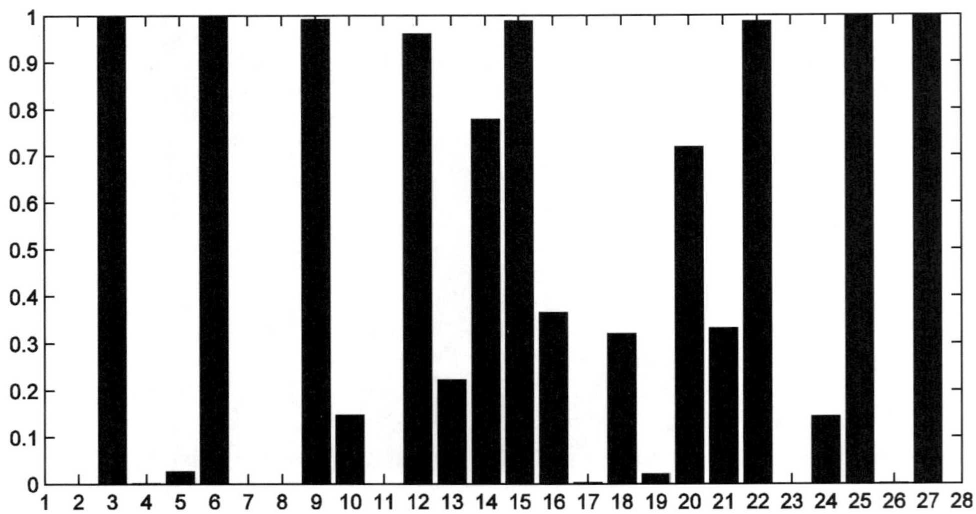

Fig. 14.2 Probability of sensor selection: EFI with stochastic Young's modulus

simulation is performed with the deterministic finite element code for the given number of samples. The natural frequencies and mode shapes of the truss bridge are considered as the output response. The standard deviation of these frequencies for different number of samples for the MCS is used to determine the number of samples required. In this case, due to the stabilization of the standard deviation, 2,000 samples are chosen.

14.3.2 Sensor Placement Under Parametric Uncertainty

Initially, the effect of uncertain parameters on the sensor configuration is studied independently. Eleven sensor locations are chosen out of the 28 DOFs available using the four OSP methods discussed in Sect. 14.2. In MCS, the cross-sectional and geometric properties vary with each sample. Therefore, the optimal locations of the 11 sensors also vary with each sample. The probabilities of the DOFs chosen as optimal sensor locations for the case of stochastic Young's modulus, and for 2,000 samples are shown in Fig. 14.2. The results for all the stochastic variables show that there are several DOFs that are always chosen as an optimal sensor location in every sample of the MCS. These sensor locations are of vital importance in the global configuration and can be called as vital sensors. The results of EFIWM and KEMRO methodologies show similar sensor configurations due to their theoretical similarities. In the case of the EFI method, DOFs 3, 6, 9, 25, and 27 have a probability of 1.0 when the Young's modulus or cross section or mass density is considered as uncertain parameter. Therefore, these DOFs form five vital sensor locations. The first 10 optimal sensor locations have much higher probabilities than the rest. However, the 11th and the next best optimal choice have only small differences between their probabilities. For the EFIWM algorithm, six sensors have a probability of 1.0 and they are vital sensors. However, in this case the variability between the remaining sensors is higher. The KEMRO algorithm shows a similar behaviour to EFIWM with seven vital sensors having a probability of 1.0 and small differences between the remaining optimal sensors locations. Finally the SEMRO algorithm shows the highest number of vital sensors, with eight having probabilities of 1.0, and has a large difference between the first 11 optimal locations and the others. The cumulative effect of the three random variables on the OSP using the four methods show the number of vital sensors has decreased in all four methodologies compared to the results with individual parameter uncertainty. Furthermore, all the algorithms show the same vital sensors locations compared to the corresponding cases of individual stochastic variables, except in the case of EFI where one of the sensors is different.

Table 14.2 Mean values of the assessment criteria for the suitability of measurement locations with parametric uncertainty

Method	(det(FIM))	(MAC rms error)	(SVD-ratio)
EFI	2.4757e-13	0.0269	2.0514
EFIWM	8.4894e-14	0.0107	1.5380
KEMRO	8.2864e-14	0.0103	1.5104
SEMRO	7.7107e-14	0.0276	2.1452

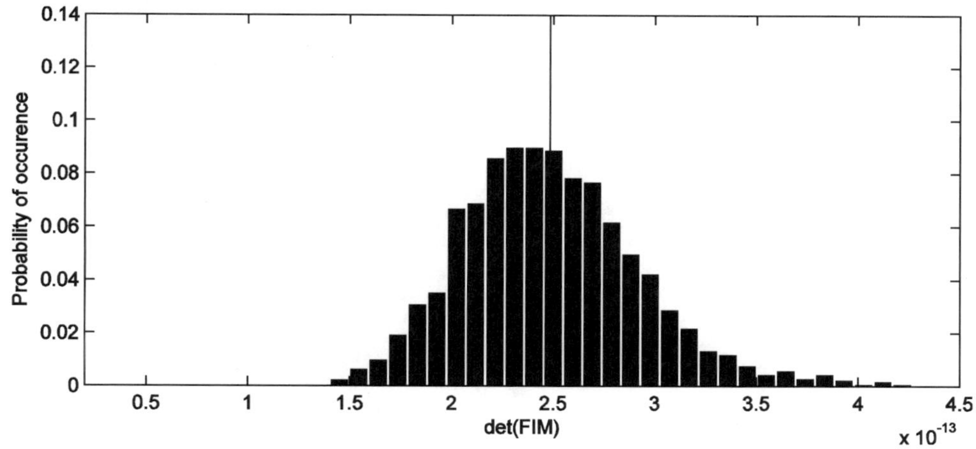

Fig. 14.3 Probability density function of det(FIM) for the EFI method

14.3.3 Assessment Criteria for OSP Methodologies

In the above section, the optimal sensor locations are chosen using different OSP methodologies, and the results show a variation in the optimal locations. Therefore, an assessment criterion should be established to compare the suitability of different sensor configurations obtained from the various OSP methodologies. In the following, three of the most important criterion are given and used in [1].

1. Determinant of Fisher information matrix (FIM).
2. RMS value of the off-diagonal terms of modal assurance criterion (MAC) matrix.
3. Singular value decomposition ratio (SVD-ratio) of the mode shape matrix.

These three assessment criteria are computed for every sample of the MCS. Table 14.2 presents the mean value of these criteria for all of the four OSP methodologies employed in the present work. As an example, histogram plotted in Fig. 14.3 represents the distribution of the assessment criteria for every OSP algorithm. In the case of the determinant of the FIM, the EFI algorithm always gives the highest value for all of the methods. This observed result is obvious if the mathematical basis of the EFI method is studied, due to the assumption of the maximization of this determinant. The evolution of the MAC rms ratio clearly illustrates that it offers the best solution in the case of the EFIWM and KEMRO algorithms. A similar result can be observed if the SVD ratio evolution is considered. Finally, it is observed that the behaviour of the EFIWM and KEMRO algorithms is quite similar whatever criterion is employed.

14.3.4 Robustness of Optimal Sensor Configurations

Measurement techniques for experimental and operational modal analysis are subjected to noise effects that contaminate the experimental data. The sensor configuration must be chosen allowing for the existence of these noise effects. In particular this determines the number of the sensors that need to be employed in the experimental tests. Furthermore, a certain level of accuracy is necessary to demonstrate the assessment of the sensor configuration. Noise effects can be incorporated into finite

Table 14.3 Mean MAC rms error for the suitability of three different measurement locations with parametric uncertainty and different noise levels

	μ (MAC rms error)			
	No noise	40 dB	30 dB	10 dB
11 sensors	0.0334	0.0377	0.0670	0.1422
14 sensors	0.0332	0.0363	0.0580	0.1131
25 sensors	0.0039	0.0064	0.0247	0.0659

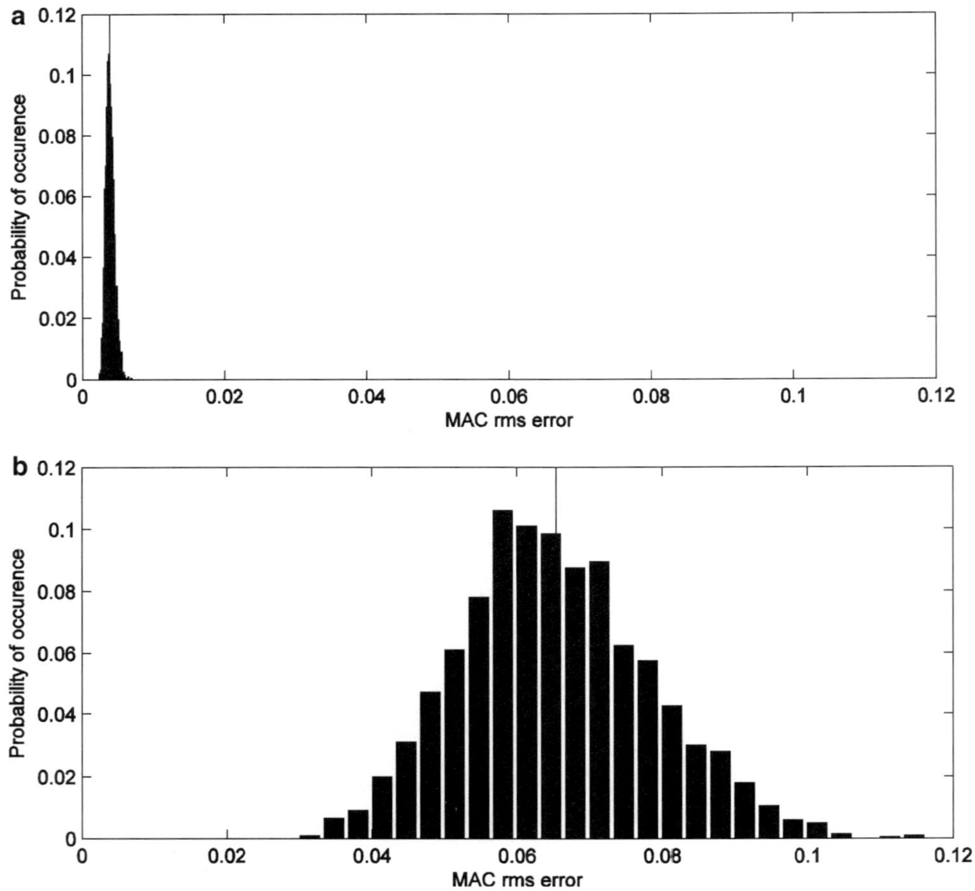

Fig. 14.4 Probability density function of the MAC rms error for 25 sensors with (**a**) no noise and (**b**) 10 dB of noise

element models numerically to simulate their consequences. The signal-to-noise-ratio (SNR) is employed as a measure of the noise level. The SNR is defined as the power ratio between a signal (meaningful information) and the background noise (unwanted signal).

$$SNR = \frac{P_{signal}}{P_{noise}} \tag{14.6}$$

In this case the noise is added to the meaningful information from the OSP objective, i.e. added to the target mode shapes. Different SNR levels are considered, from a low noise case (SNR = 40 dB) to a high noise case (SNR = 10 dB). In the ideal case of no-noise the evolution of the MAC rms error for different sensor configurations show how this value improves when the number of sensors increase. For this example, the EFI method was selected as the sensor placement algorithm and was implemented for different numbers of sensors. The MAC rms error for the best sensor locations for each number of sensor decreases when the number of sensors increases. If noise is present then the MAC rms error increases. Table 14.3 presents the mean value for the MAC rms error for different noise levels and also for different numbers of sensors. Figure 14.4 represents the MAC rms error histograms if all DOFs are measured at the same time and for different levels of SNR.

The number of sensors can also be selected if a desired level of accuracy is required. Sensors configurations become very expensive as the number of sensors increases. Furthermore, a procedure is required to estimate the exact number of sensors needed. In particular, for similar levels of accuracy measured in terms of MAC rms error, big differences could exist in the number of sensors employed. To illustrate this methodology suppose we wish to select the number of sensors between 11 and 14. The effects of the noise are also considered. In this case the EFI algorithm is applied to determine the best sensor configuration for the case of 11 and 14 sensors. After these configurations are fixed, the MCS is run with uncertain parametric variables (Young's modulus, cross section and mass density) to assess the performance in terms of MAC rms error. If noise is added the MAC rms error increases from the case of low noise. The MAC rms errors are similar for 11 and 14 sensors in the case of no noise, but show higher differences with high noise.

14.4 Conclusions

This paper investigates the influence of parametric uncertainties on the OSP methodologies for modal analysis of structures. Four OSP algorithms were used for numerical studies. A truss type bridge structure is considered as the test case to study the effects of parametric uncertainties on the OSP. Geometric (cross section) and material properties (Young's modulus and mass density) were considered as uncertain variables. Monte Carlo simulation was performed to evaluate the optimal sensor locations under uncertainty. This was propagated through a finite element model of the truss bridge structure. The natural frequencies show considerable scatter due to uncertainty. In addition, the mode shapes also show considerable variations due to uncertainty and therefore affect the optimal sensor locations. Initially, the effects of uncertainty in individual variables on the sensor configurations were examined. The numerical results show that some sensor locations were always chosen even with the uncertainty. These sensors are called vital sensors due to their high probabilities of occurrence. When the cumulative effect of uncertain variables are considered, the MCS shows similar results and retains the vital sensors of the individual uncertainty results. In order to compare the optimal sensor configurations obtained from four OSP methodologies, three assessment criteria were established. These criteria were applied to the sensor configurations identified by the four OSP algorithms considered in this work. Furthermore, due to the importance of noise in real applications, different levels of signal-to-noise-ratios were considered. The noise was incorporated into the finite element model to simulate its effects on the mode shapes of truss bridge. Three different levels of noise were considered and the corresponding sensor configurations were compared. Finally, these numerical results demonstrate that uncertainties in geometric and material properties play a significant role in the optimal sensor placement methodologies.

Acknowledgements We express special thanks to the Spanish Ministry of Education, Culture and Sport for Grant Number FPU-AP2009-3475 and to the Junta de Andalucía for the Research Project P09-TEP-5066.

References

1. Penny JET, Friswell MI, Garvey SD (1994) Automatic choice of measurement locations for dynamic testing. AIAA J 32:407–414
2. Garvey SD, Friswell MI, Penny JET (1996) Evaluation of a method for automatic selection of measurement locations based on subspace-matching. In: Proceedings of XIV international modal analysis conference (IMAC), Hyatt Regency Dearborn Hotel, Dearborn, pp 1546–1552
3. Li DS, Li HN (2006) The state of the art of sensor placement methods in structural health monitoring. In: Tomizuka M, Yun CB, Giurgiutiu V (eds). In:Proceedings of the SPiE, Smart structures and materials 2006: sensors and smart structures technologies for civil, mechanical, and aerospace systems, vol 6174, pp 1217–1227
4. Kammer DC (1991) Sensor placement for on-orbit modal identification and correlation of large space structures. J Guid Contr Dyn 14:251–259
5. Kammer DC, Peck JA (2008) Mass-weighting methods for sensor placement using sensor set expansion techniques. Mech Syst Signal Process 22:1515–1525
6. Hemez FM, Farhat C (1994) An energy based optimum sensor placement criterion and its application to structural damage detection. In: Proceedings of XII international modal analysis conference (IMAC), Ilikai Hotel, Honolulu, pp 1568–1575
7. Heo G, Wang ML, Satpathi D (1997) Optimal transducer placement for health monitoring of long span bridge. Soil Dyn Earthq Eng 16:495–502
8. Schueller GI (2007) On the treatment of uncertainties in structural mechanics and analysis. Comput Struct 85:235–243
9. Choi SK, Grandhi RV, Canfield RA (2006) Reliability-based structural design. Springer, London
10. Doebling SW, Hemez FM (2001) Overview of uncertainty assessment for structural health monitoring. In: Proceedings of the 3rd international workshop on structural health monitoring, Stanford University, Stanford
11. Murugan S, Ganguli R, Harursampath D (2008) Aeroelastic response of composite helicopter rotor with random material properties. J Aircr 45:306–322

12. Murugan S, Harursampath D, Ganguli R (2008) Material uncertainty propagation in helicopter nonlinear aeroelastic response and vibration analysis. AIAA J 46:2332–2344
13. Murugan S, Chowdhury R, Adhikari S, Friswell MI (2011) Helicopter aeroelastic analysis with spatially uncertain rotor blade properties. Aero Sci Tech 16:29–39
14. Austrell PE (2004) CALFEM: A Finite Element Toolbox : Version 3.4. Lund University (Sweden)
15. Guratzsch RF, Mahadevan S (2010) Structural health monitoring sensor placement optimization under uncertainty. AIAA J 48:1281–1289

Chapter 15
Effects of Bolted Connection on Beam Structural Modal Parameters

Zhi-Sai Ma, Li Liu, Si-Da Zhou, Di Jiang, and Yuan-Yuan He

Abstract Engineering structures are generally designed as the assembly of several parts in consideration of manufacture and structural strength, while different connections have great effects on the dynamic characteristics of the whole structure. This article firstly introduces the influence of stiffness and mass distribution on the beam's modal parameters. Then a beam-like structure consisting of four steel parts connected with bolts is presented and a series of modal testing is performed to estimate the modal parameters under free-free boundary condition. Afterwards connection location and bolt preload are selected as variables for modal testing, and the modal parameters of the beam-like structure with different connection locations and bolt preloads are compared. Finally, the finite element model of the bolted connection is updated based on the data of modal testing. The results indicate that the larger bolt preload increases the stiffness of connections, which consequently changes the modal parameters, while the influence of the bolt preload highly depends on the connection location.

Keywords Bolt preload • Bolted connection • Finite element analysis • Modal testing • Structural modal parameter

15.1 Introduction

Structural dynamic analysis plays an important role during the preliminary design of engineering structures, especially for the large-scale engineering structures, such as bridges, deployable structures and launch vehicles. Large-scale engineering structures are usually designed as the assembly of several parts in consideration of manufacture and structural strength. Generally, the structure geometry, the boundary conditions and material characteristics the mass, stiffness and damping distribution of the structure contain sufficient information to determine the system modal parameters (natural frequencies, damping factors and mode shapes) [1]. For large-span structures, different parts can be connected with each other through bolts, and the bolted connection influences the structural dynamic characteristics of the whole structure. Compared with the structure without any connections, the bolted connection changes the local stiffness of the structure. Furthermore, the contact between the bolts and different structure components causes localized non-linear stiffness and damping changes [2]. As many structures are assemblies, modal testing and the proper finite element model of the bolted connection are needed to better understand the characteristics of the bolted connection.

Effects of the bolted connection on structural modal parameters had been studied by many people. R.A. Ibrahim and C.L. Pettit provided an overview of the problems pertaining to structural dynamics with bolted joints in their review article [3]. Experimental and theoretical studies of a bolted joint excited by a torsional dynamic load were carried out by H. Ouyang [4]. Many finite element models had been accumulated for the analysis of bolted connections. Some models

Z.-S. Ma • L. Liu (✉) • S.-D. Zhou • D. Jiang • Y.-Y. He
School of Aerospace Engineering, Beijing Institute of Technology, Zhongguancun South Street 5, Haidian District, Beijing 100081, China
e-mail: zhisai.ma@gmail.com; liuli@bit.edu.cn

A. Cunha (ed.), *Topics in Dynamics of Bridges, Volume 3: Proceedings of the 31st IMAC, A Conference on Structural Dynamics, 2013*, Conference Proceedings of the Society for Experimental Mechanics Series 38,
DOI 10.1007/978-1-4614-6519-5_15, © The Society for Experimental Mechanics, Inc. 2013

were employed to structural static analysis; others were used to analyze structural dynamic problems, and it focuses on the latter here. For instance, linear dynamic behavior of bolted joint was modelled and its parameters were identified using experimental observations by H. Ahmadian [5]. A simple, reliable and feasible finite element model with three-dimensional solid elements about blind hole bolted connection was proposed to obtain bolt pre-tensional effect in reference [6]. The identification of bolted lap joints parameters in assembled structures was done by H. Ahmadian and H. Jalali [7]. Nevertheless, few suitable models and simplified analysis methods can be used for some specific problem. Therefore, this article aims to offer the reader a good understanding for investigating the dynamics of the beam-like structure with bolted connections.

In the present article, the foundational theory on structural dynamics is briefly reviewed, along with the influence of stiffness and mass distribution on the beam's modal parameters (Sect. 15.2). Modal testing on the beam-like structure with different bolted connections is conducted; and effects of connection location and bolt preload on its modal parameters are also obtained (Sect. 15.3). A simplified but reliable finite element model of the bolted connection is built, and the spring stiffness of the finite element model is optimized based on the data of modal testing (Sect. 15.4). The conclusions of the study are finally presented (Sect. 15.5).

15.2 Beam Flexure

The formal mathematical procedure for considering the behavior of an infinite number of connected points is by means of differential equations in which the position coordinates are taken as independent variables [8]. Inasmuch as time is also an independent variable in a dynamic response problem, the formulation of the equations of motion in this way leads to partial differential equations. Here, attention will be limited to one dimensional structures such as beam-type systems which may have variable mass, stiffness and damping properties along their elastic axes. The partial differential equations of these systems involve only two independent variables: time and distance along the elastic axis of each component member.

15.2.1 Partial Differential Equation of Motion

For a straight, nonuniform beam, the significant physical properties of this beam are assumed to be the flexural stiffness $EI(x)$ and the mass per unit length $m(x)$, both of which may vary arbitrarily with position x along the span L. The transverse loading $p(x, t)$ is assumed to vary arbitrarily with position and time, and the transverse-displacement response $v(x, t)$ is also a function of these variables. For arbitrary boundary conditions, the partial differential equation of motion for the elementary case of beam flexure [8] can be obtained as below

$$\frac{\partial^2}{\partial x^2}\left[EI(x)\frac{\partial^2 v(x,t)}{\partial x^2}\right] + m(x)\frac{\partial^2 v(x,t)}{\partial t^2} = p(x,t)$$

Of course, the solution of this equation must satisfy the prescribed boundary conditions at $x = 0$ and $x = L$. The free-vibration equation of motion for this system is

$$\frac{\partial^2}{\partial x^2}\left[EI(x)\frac{\partial^2 v(x,t)}{\partial x^2}\right] + m(x)\frac{\partial^2 v(x,t)}{\partial t^2} = 0$$

It can be found that the flexural stiffness $EI(x)$ and the mass per unit length $m(x)$ determine the modal parameters of the structure. The natural frequencies, along with their mode shapes, can be obtained through solving this free-vibration equation of motion. However, the solutions of the equations of motion for most complex systems generally can be obtained only by numerical means.

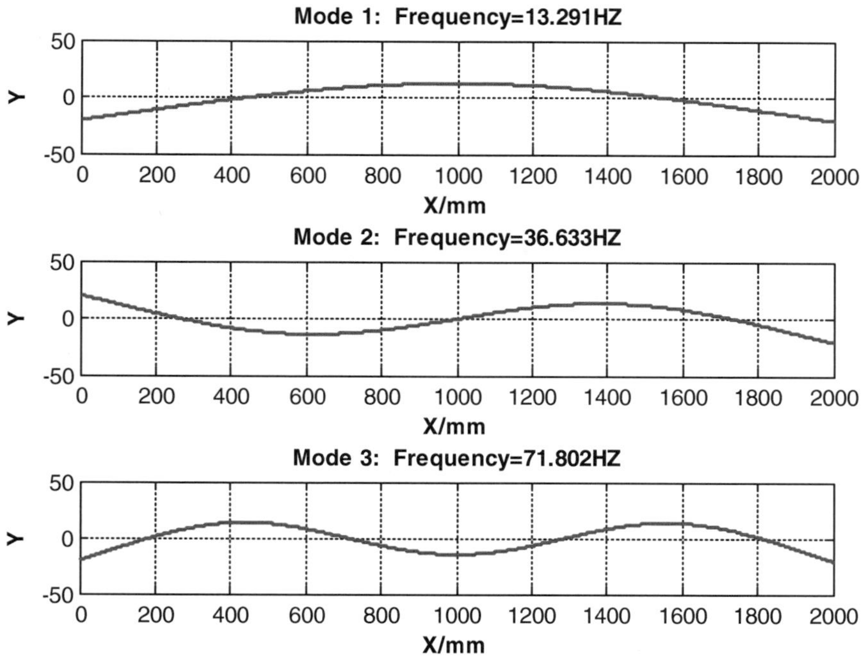

Fig. 15.1 Natural frequencies and mode shapes of benchmark beam

15.2.2 *Benchmark Beam*

As shown in the free-vibration equation of motion, the flexural stiffness *EI(x)* and the mass per unit length *m(x)* determine the natural frequencies and the corresponding mode shapes of the structure. For a steel beam with dimensions 2,000 × 60 × 10 mm (L × W × H), its mass density is 7,850 kg/m^3, Poisson's ratio is 0.33 and Young's modulus is 210 GPa. The flexural stiffness *EI(x)* and the mass per unit length *m(x)* equal to constants *EI* and *m*, respectively. The first three natural frequencies and the corresponding mode shapes are obtained for this benchmark beam under free-free boundary condition, as depicted in Fig. 15.1.

15.2.3 *Effects of Stiffness and Mass Distribution on Modal Parameters*

In order to discuss the effects of stiffness and mass distribution on the beam's modal parameters, the flexural stiffness *EI(x)* and the mass per unit length *m(x)* may vary arbitrarily with position x along the span of the beam. For a beam with dimensions 2,000 × 60 × 10 mm (L × W × H), its flexural stiffness *EI(x)* or the mass per unit length *m(x)* at specific position ($x = x_0$) are different from the others during the finite element analysis, as follows,

$$EI(x) = \begin{cases} E_1 I & x \neq x_0 \\ E_2 I & x = x_0 \end{cases}, \quad m(x) = \begin{cases} \rho_1 A & x \neq x_0 \\ \rho_2 A & x = x_0 \end{cases} \quad (x \in [0, L])$$

in which A is the area of the beam cross-section.

On the one hand, only the effects of flexural stiffness distribution on the natural frequencies are considered, that means the influence of the mass distribution is ignored during this process through making $\rho_2 = \rho_1$, If x_0 takes all the values within the interval $[0, L]$ in turn, then we can find the effects of flexural stiffness distribution on the natural frequencies. As depicted in Fig. 15.2, the vertical axis is the value of the natural frequency, and the horizontal axis is the value of x_0. The dotted line represents the frequency of the benchmark beam with constant EI and m ($E = 210GPa$, $\rho = 7{,}850kg/m^3$), the solid line represents the frequency of the stiffness-varying beam ($E_2 = 1.5E_1 = 1.5E$, $\rho_2 = \rho_1 = \rho$). It is obvious that natural frequencies of the structure can be increased effectively through increasing the local stiffness of the peak positions in mode shapes shown in Fig. 15.1, and the change of local stiffness at nodal points for specific mode shape has little influence on the corresponding frequency.

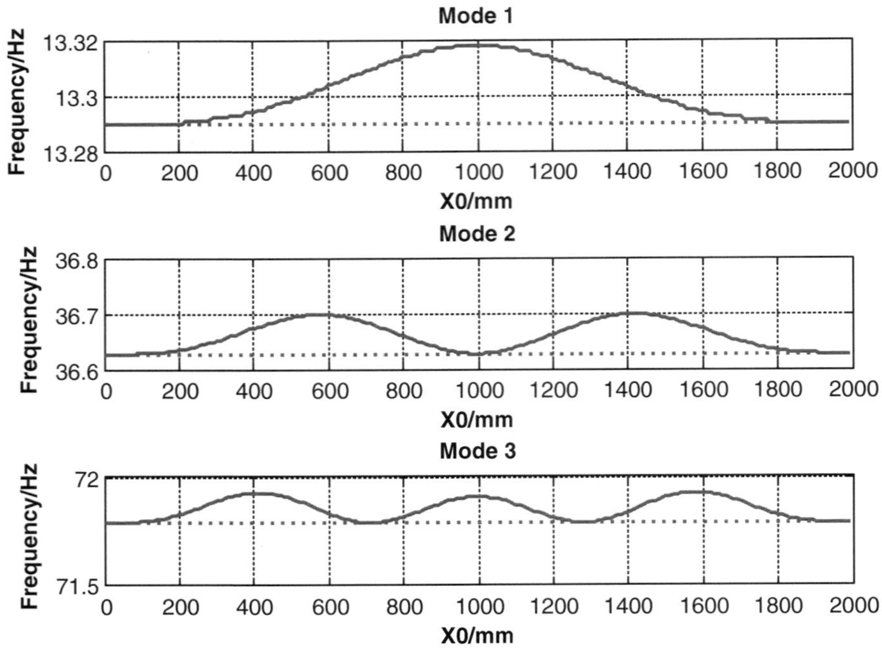

Fig. 15.2 Effects of stiffness distribution on natural frequencies

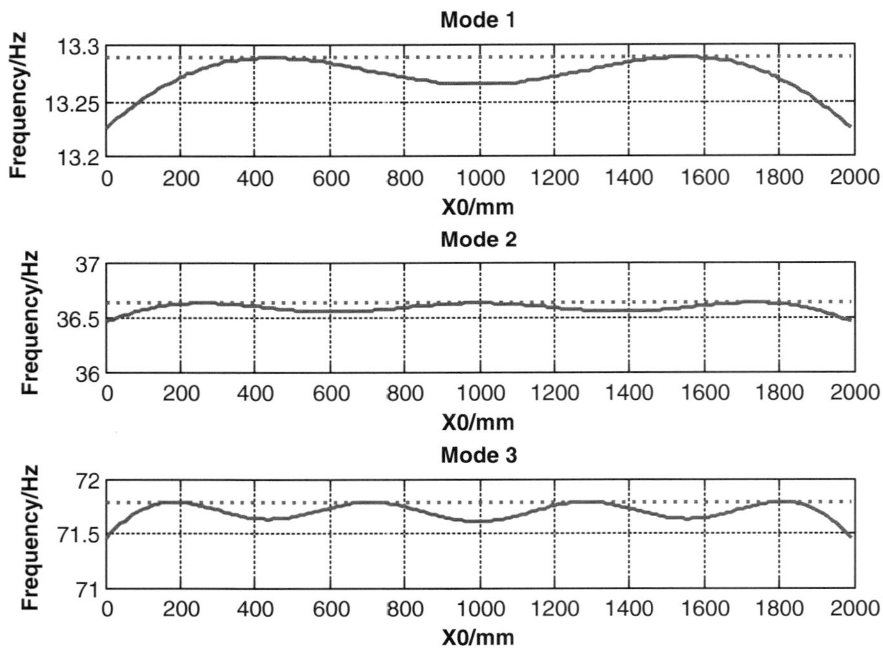

Fig. 15.3 Effects of mass distribution on natural frequencies

On the other hand, the effects of the mass distribution on the natural frequencies are considered, and the influence of the stiffness distribution is ignored ($E_2 = E_1$). As depicted in Fig. 15.3, the vertical axis is the value of the natural frequency, and the horizontal axis is the value of x_0. The dotted line represents the frequency of the benchmark beam with constants EI and m ($E = 210GPa, \rho = 7{,}850kg/m^3$), the solid line represents the frequency of the mass-varying beam ($E_2 = E_1 = E, \rho_2 = 1.5\ \rho_1 = 1.5\ \rho$). It can be found that natural frequencies of the structure will descend obviously when the larger mass element locates at the peak positions of the mode shapes shown in Fig. 15.1, and the change of local mass at nodal points for specific mode shape has little influence on the corresponding frequency. It should be noted that the mass distribution at the ends of the beam has great effects on the natural frequencies of the beam under free-free boundary condition.

The natural frequencies of the beam are influenced by the flexural stiffness $EI(x)$ and the mass per unit length $m(x)$. Generally, the larger local flexural stiffness or mass per unit length will lead to the increment or descent of the natural frequencies, respectively. Furthermore, any changes at nodal points for specific mode shape have little influence on the corresponding frequency and this characteristic can be utilized in the design of engineering structures.

15.3 Modal Testing on Beam-Like Structure

Firstly, a beam-like structure consisting of four steel parts connected with bolts is introduced, and then the experimental set-up for the modal testing is introduced. Finally, the modal testing for this beam-like structure with different bolted connections is conducted to study the effects of the connection location and the bolt preload on its modal parameters.

15.3.1 Beam-Like Structure

The beam-like structure is composed of four steel parts and three joints can be found on it. The dimensions of the beam-like structure are 2,000 × 60 × 10 mm (L × W × H) with each part of dimensions 500 × 60 × 10 mm (L × W × H). Different parts are connected with each other through eight bolts, as shown in Fig. 15.4.

In order to allow testing this structure, there will always exist a connection between the structure and the environment. The free-free condition means no connections with environment should exist. In practice this condition is approximated by hanging the test structure on very soft springs [1]. To minimize the influence of the suspension, the beam-like structure is suspended in its lateral direction so that the bending modes will appear in a plane which is perpendicular to the suspension direction, as shown in Fig. 15.4.

15.3.2 Modal Testing Set-Up

Modal testing set-up consists of test structure, exciter systems, force and motion transducers, measurement and analysis systems, and suspension systems. The hammer impact is selected as the vibration force to be applied to the test structure, which is PCB^{TM} D086C03 excitation hammer (sensitivity 2.25 mV/N, resonance frequency $\geq$ 22 kHz). The motion

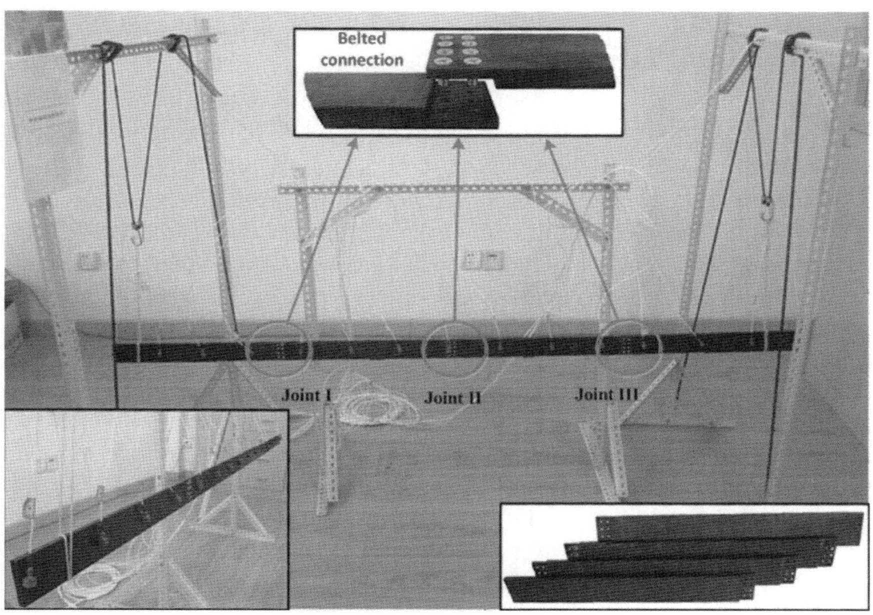

Fig. 15.4 Schematic diagram of the beam-like structure

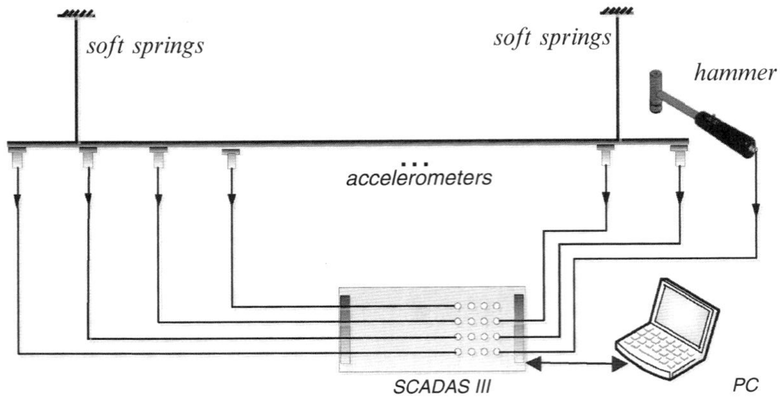

Fig. 15.5 Schematic diagram of the modal testing set-up

Table 15.1 Groups of experimental data

<table>
<tr><th>Group / Joint</th><th colspan="4">Group A (Nm)</th><th colspan="4">Group B (Nm)</th><th colspan="4">Group C (Nm)</th></tr>
<tr><td>Joint I&III</td><td rowspan="2">1.0</td><td rowspan="2">1.5</td><td rowspan="2">2.0</td><td rowspan="2">2.5</td><td colspan="4">1.0–2.5</td><td>1.0</td><td>1.5</td><td>2.0</td><td>2.5</td></tr>
<tr><td>Joint II</td><td>1.0</td><td>1.5</td><td>2.0</td><td>2.5</td><td colspan="4">1.0–2.5</td></tr>
</table>

transducer is the PCB™ 333B30 accelerometer (frequency range 0.5–3,000 Hz, sensitivity 10.2 mV/(m/s^2) and resonance frequency $\geq$ 40 kHz). Measurement and acquisition module is LMS™ SCADAS III system (16 channels). The schematic diagram of the modal testing set-up is depicted in Fig. 15.5.

The bolt preload should be accurately controlled to obtain the effects of the bolt preload on the structural dynamic characteristics of the test structure. The bolt preload is usually measured by the torque meter, and HIOS HP-100 torque meter (range 0.15–10 Nm, accuracy $\pm$0.5 %) is used here. The standard torque-limiting screwdriver (range 0.06–4.06 Nm, accuracy $\pm$6 %) from Torqueleader™ is used to tighten the bolt, and the set torque of the screwdriver is controlled by the HP-100 torque meter. In this way, we know exactly the value of the preload on every bolt in each test, and the effects of the bolt preload on the beam-like structure can be analyzed through the modal testing.

15.3.3 Modal Testing

Each modal analysis test should start with a check of the linear dynamics behavior of the test structure [1], while sometimes modal testing can be used to analyze those structures with weak nonlinearities. The contact between the bolts and different components leads to nonlinearities in structural dynamics, but these localized nonlinearities in structural dynamics are ignored during the modal testing for this beam-like structure. In other words, linear analysis method is utilized to simulate the dynamic characteristics of the structure with weak and localized nonlinearities here.

Modal testing for the beam-like structure is conducted to obtain the modal parameters of the structure with different connection locations and bolt preloads under free-free boundary condition. In the modal testing, the bolt preload at different joints can be different but all the eight bolts at each joint have the same preload. The three joints are named as Joint I, Joint II and Joint III form left to right along the beam-like structure, as depicted in Fig. 15.4. The value of the bolt preload in each joint can be set as 1.0, 1.5, 2.0 or 2.5 Nm.

The effects of the bolted connection are studied through the comparison between these modal parameters form a series of modal testing with different connection locations and bolt preloads. In consideration of the symmetry of this beam-like structure, the bolts at Joint I and III are taken as the first variable, and the bolts at Joint II are selected as another variable. In order to easily process the data of the modal testing and find the effects of the bolt preload on the modal parameters, all the experimental data is allocated into three groups, that is A, B and C. Table 15.1 depicts the bolt preload at every joint in each group during the modal testing.

15.3.4 Results

15.3.4.1 Group A

For Group A shown in Table 15.1, all the bolts at every joint have the same preload in each test, such as 1.0, 1.5, 2.0 and 2.5 Nm, respectively. In this way, we can find how the values of bolt preload affect the modal parameters of the beam-like structure. Figure 15.6 depicts the natural frequencies of the first six modes for this beam-like structure under different bolt preloads, and Fig. 15.7 depicts the corresponding damping ratio of the first six modes.

As shown in Fig. 15.6, the vertical axis is the value of the natural frequencies, and the horizontal axis is the value of bolt preload. It can be found that the larger bolt preload increases the natural frequencies of the beam-like structure, especially for the structure under lower bolt preload.

As shown in Fig. 15.7, the vertical axis is the value of the damping ratio, and the horizontal axis is the value of bolt preload. It can be found that the larger bolt preload will lower the damping ratio of the beam-like structure; however, this trend is not very obvious. The effects of the bolted connection on the damping ratio of this beam-like structure are not focused on in this article, because the accuracy of damping ratio estimation is not as good as natural frequency estimation.

15.3.4.2 Group B

For Group B shown in Table 15.1, the value of bolt preload at Joint II keep constant in each modal test, that is, 1.0, 1.5, 2.0 and 2.5 Nm, respectively; while the bolts at Joint I and III have the same but changing preload from 1.0 to 2.5 Nm. The effects of the bolt preload at Joint I and III on the modal parameters are studied. Figure 15.8 depicts the natural frequencies of the first six modes for this beam-like structure affected by bolt preload at Joint I and III.

As shown in Fig. 15.8, the vertical axis is the value of the natural frequencies, and the horizontal axis is the value of bolt preload at Joint II. The four lines in each figure represent the natural frequencies of the structure under different bolt preloads at Joint I and III, such as, 1.0, 1.5, 2.0 and 2.5 Nm. For the mode 1 and mode 5 shown in Fig. 15.8, it can be found that the larger bolt preload at Joint II leads to larger natural frequencies, but the larger bolt preload at Joint I and III have little influence on the natural frequencies, because the locations of Joint I and III are close to the nodal points of mode 1 and mode 5. For the even modes shown in Fig. 15.8, it is obvious that the larger bolt preload at Joint I and III can increase the natural

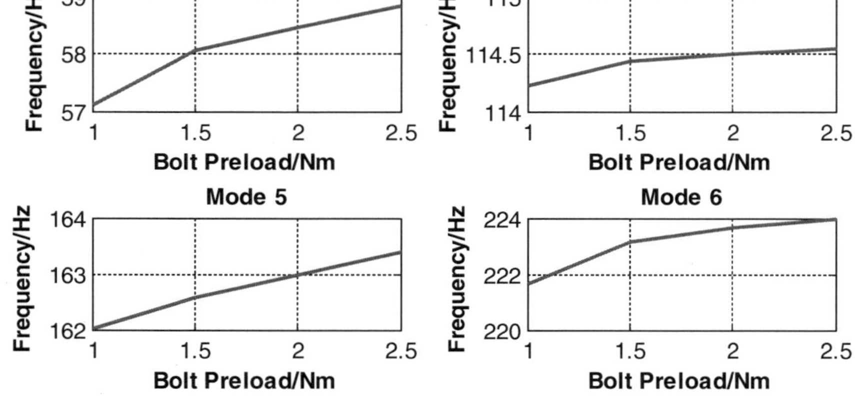

Fig. 15.6 Natural frequency affected by the value of bolt preload

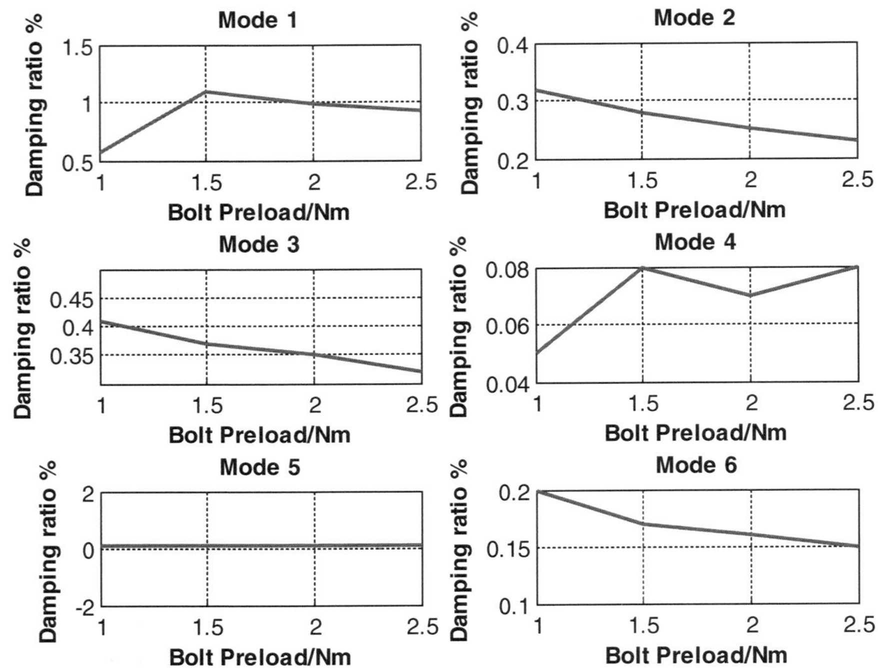

Fig. 15.7 Damping ratio affected by the value of bolt preload

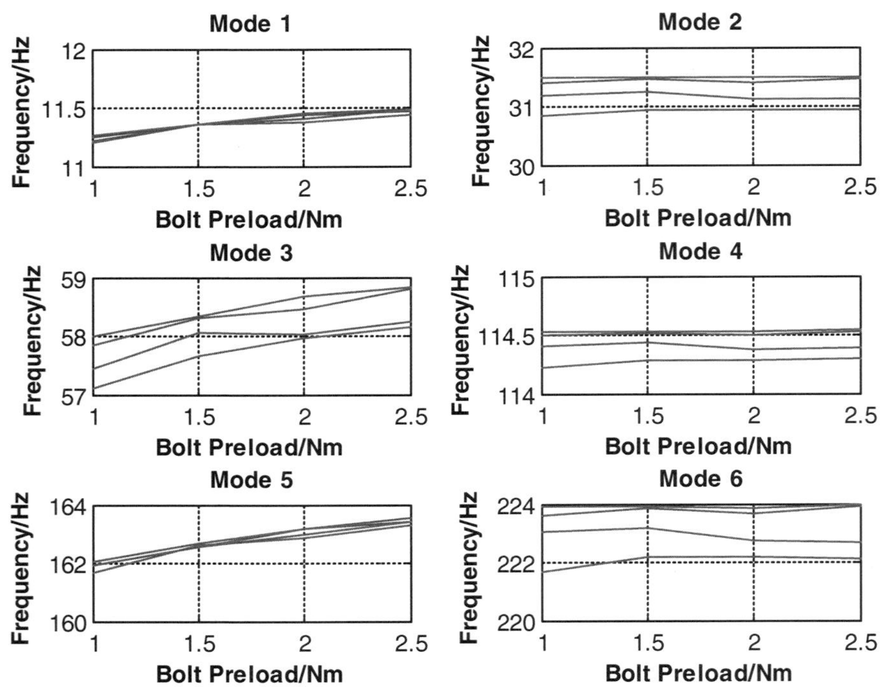

Fig. 15.8 Natural frequency affected by bolt preload at Joint I and III

frequencies of the structure under lower bolt preload. It can be also found that the changing of the bolt preload at Joint II has little influence on the even modes and the reason will be presented later in Group C. All the three joints locate away from the nodal points of mode 3, therefore, only the similar conclusion as Group A can be obtained based on the data from mode 3, that is, the larger bolt preload increases the natural frequencies of the structure with bolted connections.

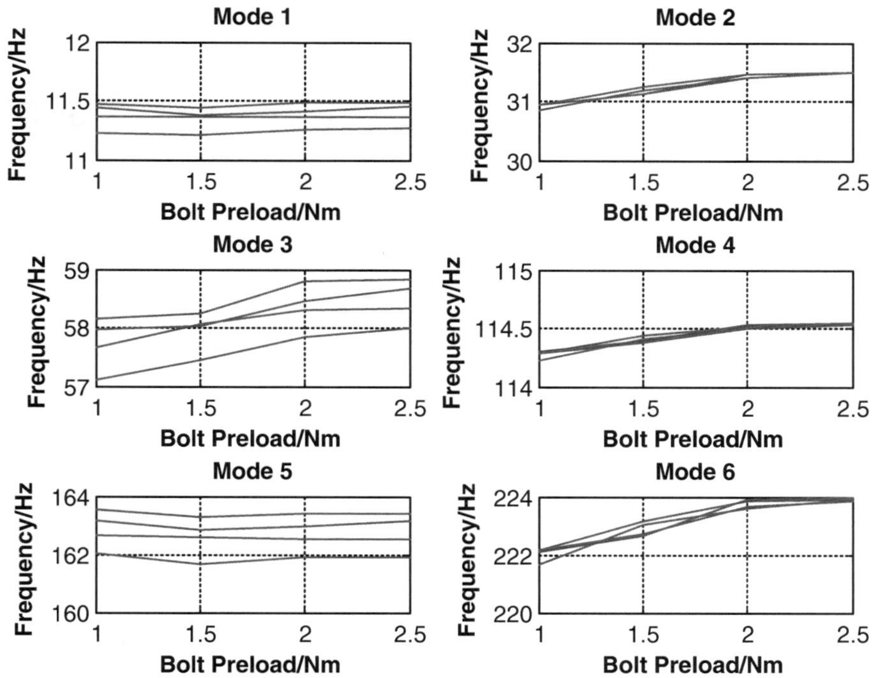

Fig. 15.9 Natural frequency affected by bolt preload at Joint II

15.3.4.3 Group C

For Group C shown in Table 15.1, the value of bolt preload at Joint I and III keep constant in each modal test, that is, 1.0, 1.5, 2.0 and 2.5 Nm, respectively; while the bolts at Joint II have the changing preload from 1.0 to 2.5 Nm. The effects of the bolt preload at Joint II on the modal parameters are studied. Figure 15.9 depicts the natural frequencies of the first six modes for this beam-like structure affected by bolt preload at Joint II.

As shown in Fig. 15.9, the vertical axis is the value of the natural frequencies, and the horizontal axis is the value of bolt preload at Joint I and III, The four lines in each figure represent the natural frequencies of the structure under different bolt preloads at Joint II, such as, 1.0, 1.5, 2.0 and 2.5 Nm. For the even modes shown in Fig. 15.9, it is obvious that the increment of the bolt preload at Joint II cannot increase the natural frequencies of these even modes, because Joint II locates on the midpoint of the structure, which is the nodal point of even modes for symmetrical structures. It can be also found that the increment of the bolt preload at Joint I and III can increase the natural frequencies of even modes. For the odd modes shown in Fig. 15.9, the larger bolt preload at Joint II increases these natural frequencies of the structure. However, the increment of the bolt preload at Joint I and III cannot increase the natural frequencies of the mode 1 and mode 5, because the locations of Joint I and III are close to the nodal points of mode 1 and mode 5.

Above results indicate that the bolt preload determines the stiffness of the bolted connection, which influences the modal parameters of the structure. It also validates the conclusions obtained from the numerical analysis in Sect. 15.2, i.e. the larger stiffness leads to larger natural frequencies. Furthermore, any changes at the nodal points of specific mode shape have little influence on the corresponding frequency. These conclusions can be extended to other structures and these characteristics can be utilized to obtain desirable modal parameters of the engineering structures during their design process.

15.4 Finite Element Model of Bolted Connection

Modal testing has been widely used to estimate the modal parameters of the test structure, while the cost of modal testing is much more than the numerical analysis, such as finite element analysis (FEA). Besides, it is impracticable to conduct the modal testing for some engineering structures during their practical applications. Therefore, it is necessary to setup the finite element model (FEM) of the bolted connection, and the finite element analysis will be helpful to better understand the characteristics of the bolted connection.

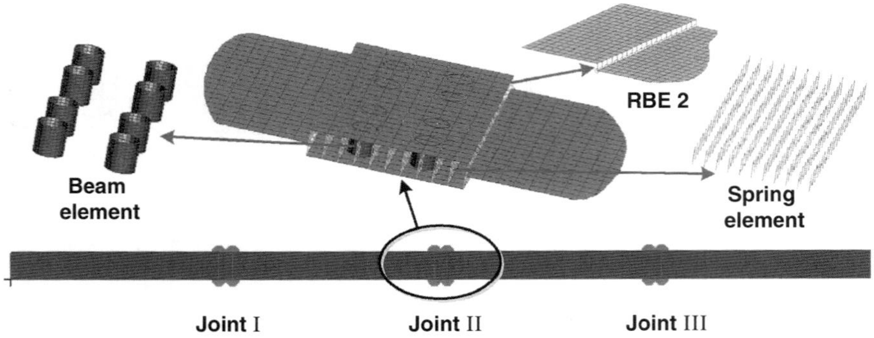

Fig. 15.10 Finite element model of bolted connection

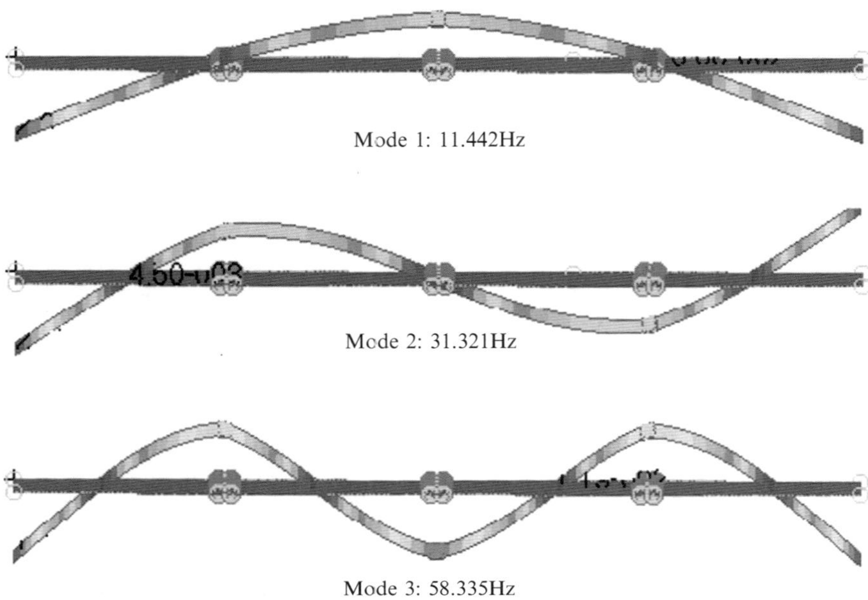

Fig. 15.11 First three modes of the structure from finite element analysis

15.4.1 *Finite Element Model*

Midsurface shell elements are used to build the finite element model of the beam-like structure with four steel parts. For each steel part, the midsurface shell elements locate in different planes when the thickness of the part varies. Therefore, the RBE2 from Multi-Point Constraint (MPC) is used to connect those elements in same steel part, but in different planes, as shown in Fig. 15.10. For the connection locations, the beam element is set up to simulate the bolt. The cross-section, material and properties of the beam element are set based on those of the bolts used in the modal testing. Spring elements in two orthogonal directions are used to simulate the interaction between different steel parts, such as extrusion and friction, as shown in Fig. 15.10.

Firstly, the first three modes of the beam-like structure are obtained without considering the stiffness of spring elements, which means the stiffness of spring elements is set to zero and different steel parts are connected with each other through beam elements only. The first three natural frequencies and the corresponding mode shapes are depicted in Fig. 15.11. Compared with the data of modal testing in Group A, the finite element model without spring elements has the similar natural frequencies with the structure, whose bolt preload is 2.0 Nm for all the joints. In order to obtain a proper finite element model of the beam-like structure with larger bolt preload, the stiffness of spring elements must be considered. Model updating is adopted to set up the proper finite element model of the bolted connection, and related theories can be found in reference [9, 10].

15.4.2 *Optimization Model*

Modal parameters of the finite element model are influenced by the stiffness of spring elements, and different spring stiffness can be used to simulate different bolt preloads. It is necessary to select the appropriate stiffness values for these spring elements to obtain the reasonable finite element model of the bolted connection. Here, the optimization method based on genetic algorithm (GA) is adopted to determine the stiffness of spring elements. Considering the characteristics of above finite element model, the bolted connection optimization model for the beam-like structure is set up as follows,

$$\begin{cases} min \quad f(k_x, k_y) = \sqrt{\sum_{i=1}^{3} c_i \left(\frac{f_i}{F_i} - 1 \right)^2} \\ s.t. \quad k_x \geq 0, \quad k_y \geq 0 \\ 0.95F_i \leq f_i \leq 1.05F_i \\ f_1 \leq f_2 \leq f_3 \end{cases}$$

The stiffness of spring elements k_x and k_y are used to simulate the friction and extrusion interaction between different steel parts, respectively. Here, k_x and k_y are selected as the variables during the optimization process. f_1, f_2 and f_3 are the first three natural frequencies of the beam-like structure form finite element analysis and each of them is a function of k_x and k_y. F_1, F_2 and F_3 are the first three natural frequencies of the beam-like structure obtained from modal testing. c_i is the weight coefficient of different natural frequencies. The objective function is to minimize the difference of the first three natural frequencies between finite element analysis and model testing. The constraint is that each natural frequency from finite element analysis should converge to no more than a five percent (± 5 %) difference with the corresponding natural frequency from modal testing and $f_1 < f_2 < f_3$ must be satisfied during the optimization process.

15.4.3 *Optimization Results*

In this section, the first three natural frequencies of the beam-like structure with 2.5 Nm bolt preload are used to optimize the stiffness of spring elements at the joints, i.e. $F_1 = 11.488$ Hz, $F_2 = 31.503$ Hz, $F_3 = 58.835$ Hz. The weight coefficient is selected as $c_1 = c_2 = c_3 = 1$ to balance the influence of different natural frequencies on the objective function. Genetic algorithm is used to find the optimal value of the spring stiffness k_x and k_y for the structure with 2.5 Nm bolt preload. With the optimal $k_x = 1.004$ N/mm and $k_y = 334.919$ N/mm, the first three natural frequencies of the beam-like structure obtained from finite element analysis are $f_1 = 11.520$ Hz, $f_2 = 31.502$ Hz, and $f_3 = 58.768$ Hz, which are very close to the data of modal testing. Other natural frequencies can also be obtained based on the finite element model with the optimal spring stiffness, for example, $f_4 = 116.170$ Hz, $f_5 = 165.550$ Hz, and $f_6 = 224.990$ Hz, which also approach to those natural frequencies from modal testing ($F_4 = 114.544$ Hz, $F_5 = 165.395$ Hz, $F_6 = 223.941$ Hz).

A simple and reliable finite element model of the bolted connection is built through selecting proper stiffness of spring elements at the joints, and this method can be used to obtain much more modal parameters of the beam-like structure than modal testing. In a word, the reliable finite element model of the bolted connection can be obtained through model updating and the modal parameters of the structure under arbitrary bolt preload can be analyzed by numerical means, which make the design of engineering structures more efficient and economic.

15.5 Conclusions

The influence of the stiffness and mass distribution on the beam's modal parameters is analyzed through numerical means firstly, that is, the larger local stiffness leads to the increment of the natural frequencies, and any changes at the nodal points of specific mode shape have little influence on the corresponding frequency. Afterwards a series of modal testing is conducted to estimate the modal parameters of the beam-like structure with four steel parts connected with each other through bolts, and the results of the modal testing validate the conclusions from numerical analysis before. For the beam-like structure with three bolted connections, the larger bolt preload increases the stiffness of connections, which changes the structural modal parameters, but this influence of the bolt preload highly depends on the connection location. Finally, a simple and reliable finite element model of bolted connection is built based on the data from modal testing, which makes the design of structures with bolted connections much more efficient and economic. In a word, a good design of the bolted connection is helpful to obtain the desirable dynamic characteristics of the whole structure.

References

1. Heylen W, Lammens S, Sas P (2007) Modal analysis theory and testing. Katholieke University Leuven, Leuven
2. Song Y, Hartwigsen CJ et al (2004) Simulation of dynamics of beam structures with bolted joints using adjusted Iwan beam elements. J Sound Vib 273:249–276
3. Ibrahim RA, Pettit CL (2005) Uncertainties and dynamic problems of bolted joints and other fasteners. J Sound Vib 279(3–5):857–936
4. Ouyang H, Oldfield MJ et al (2006) Experimental and theoretical studies of a bolted joint excited by a torsional dynamic load. Int J Mech Sci 48:1447–1455
5. Ahmadian H, Ebrahimi M et al (2002) Identification of bolted-joint interface models. In: Proceedings of ISMA 2002: International Conference on Noise and Vibration Engineering, Katholieke University, Leuven, Belgium, pp. 1741–1747
6. Xiang Zheng, Weiming Xia (2009) Numerical simulation of blind hole bolt connection with 3-D finite element approach. In: Proceedings of the Second International Conference on Information and Computing Science, ICIC 2009, Manchester, vol 4, pp.164–169
7. Ahmadian H, Jalali H (2007) Identification of bolted lap joints parameters in assembled structures. Mech Syst Signal Process 21:1041–1050
8. Clough R, Penzien J (1995) Dynamics of structures. Computers & Structures, Berkeley
9. Mottershead JE, Friswell MI (1993) Model updating in structural dynamics: a survey. J Sound Vib 167(2):347–375
10. Chen JC, Garba JA (1980) Analytical model improvement using modal test results. J AIAA 18(6):684–690

Printed by Printforce, the Netherlands